GOD
Does Not Play Dice

GOD
Does Not Play Dice

The Fulfillment of Einstein's Quest
for Law and Order in Nature

2nd Edition

David A. Shiang

PUSTAK MAHAL®
Delhi • Mumbai • Patna • Hyderabad • Bangalore • London

Publishers
Pustak Mahal®, Delhi
J-3/16 , Daryaganj, New Delhi-110002
☎ 23276539, 23272783, 23272784 • *Fax:* 011-23260518
E-mail: info@pustakmahal.com • *Website:* www.pustakmahal.com

London Office
51, Severn Crescents, Slough, Berkshire, SL 38 UU, England
E-mail: pustakmahaluk@pustakmahal.com

Sales Centre
10-B, Netaji Subhash Marg, Daryaganj, New Delhi-110002
☎ 23268292, 23268293, 23279900 • *Fax:* 011-23280567
E-mail: rapidexdelhi@indiatimes.com

Branch Offices
Bangalore: ☎ 22234025
E-mail: pmblr@sancharnet.in • pustak@sancharnet.in
Mumbai: ☎ 22010941
E-mail: rapidex@bom5.vsnl.net.in
Patna: ☎ 3294193 • *Telefax:* 0612-2302719
E-mail: rapidexptn@rediffmail.com
Hyderabad: *Telefax:* 040-24737290
E-mail: pustakmahalhyd@yahoo.co.in

Copyright © 2008 by David A. Shiang

Copyright in India © 2008 by Pustak Mahal, India

Published in India in arrangement with David A. Shiang

1st Indian Edition : February 2008
2nd Indian Edition : September 2008

ISBN: 978-81-223-1015-3

NOTICE

All rights reserved. No part of this book may be reproduced by any mechanical, photographic, or electronic process, or in the form of a phonographic recording, nor may it be stored in a retrieval system, transmitted, or otherwise be copied for public or private use—other than for "fair use" as brief quotations embodied in articles and reviews without prior written permission of the publisher.
This publication is designed to provide accurate and authoritative information in regard to the subject matter covered. It is sold with the understanding that the publisher is not engaged in rendering legal, accounting, or other professional service. If legal advice or other assistance is required, the services of a competent professional person should be sought.

Printed at : Param Offsetters, Okhla, New Delhi-110020

For Helen

I want to know how God created this world. I am not interested in this or that phenomenon, in the spectrum of this or that element. I want to know His thoughts. The rest are details.

Albert Einstein

And what you thought you came for
Is only a shell, a husk of meaning
From which the purpose breaks only when it is fulfilled
If at all. Either you had no purpose
Or the purpose is beyond the end you figured
And is altered in fulfilment.

T.S. Eliot

Contents

Foreword	i
1. Introduction	1
2. Beginning the Journey	7
3. A Certain World Beyond Science	19
4. The Solution to the Free Will Problem	27
5. Cracking Nature's Code of Law and Order	45
6. Quantum Leaps of Faith	65
7. Sagan's Baloney Detection Kit Applied	83
8. Why Darwinism is Defunct	105
Appendix A Observations on the Direction of Time	111
Appendix B Einstein, God, and the Number One Musketeer	115
Notes	119
Index	129
Acknowledgements	133

Foreword

I first stumbled upon David A. Shiang's very obscure, if not now, very rare first book, *On the Absence of Disorder in Nature*, on the floor all by itself at Powell's Books in Portland, Oregon quite late on a cold and rainy night in the early 1980's. I bent down to pick it up. When I cracked the pages, the text was so shocking and strange to my Stanford scientific training that I had to read and re-read the book forty or fifty times forcing me to go back to the text literally hundreds of times since then. I subsequently included it in my pantheon of favorite books. And the book is only 29 pages long. His second book, this one, is longer and just as exciting to read. It has taken me over 20 years for Shiang's new way of thinking to finally become integrated with my own preconceptions and scientific stereotypes. Perhaps the readers of this book will not take so long to understand Shiang as I did.

Shiang elegantly attacks and charmingly demolishes, randomness, atheism, Darwin and natural selection. And this does not exhaust the list. His lucidity and logic are breathtakingly devastating. He is not afraid to defend the mind of God, either. He is the most original and wildest thinker in the philosophy and psychology of science today.

Shiang has been laboring in obscurity since the late 1970's on his discoveries regarding the nature of the universe and the erroneous assumptions the scientists of our era are making regarding chaos, randomness, and God. Shiang also refers to God as "the gold mine of consciousness."

Shiang acknowledges he's been encouraged by Karl Popper, David Bohm and others to continue development of his deductions. His scientific arguments and the ways he argues them are original in scientific thought. This book is a culmination of decades of courageous and frustrating effort going against the grain of popular scientists who now ride the wave of the trendy new atheism and a universe of randomness. The book sales of these scientists are in the millions of copies. It's quite common and lucrative now to write and lecture about how God is nonexistent or a psychopath. And of course there is always the criticism about God's alleged poor craftsmanship in the universe. As if God had only him/herself to blame for all that ails our planet!

In opposition to the trendy belief in atheism and chaos, Shiang step-by-step and in elegant detail uncovers the secret of a surprisingly deep ignorance shared among today's most popular scientists. Some of the scientists have actually referred to themselves as the "brights." As one bright indicated, brights do not believe in God.

I think Shiang will outlast the trendy brights, especially on the question of randomness. After disproving something proponents of natural selection and most other lay people just take for granted – namely, that there have actually been empirical experiments that have proven the existence of randomness – he cites recent quotes of some of our premier scientific minds showing they irrationally believe in that powerful illusion. They have to believe it; otherwise, they would all have quite little to talk about.

Moreover, Shiang shows they cling to a universe full of randomness out of a need to be in control. It's a bit frightening to watch Shiang's mind explicitly illuminate how the brights we have been listening to deserve even less credibility than the God they say is a delusion. For this reason, Shiang is sure to be vilified and this book is sure to be attacked for decades. This book is that powerful.

Over the past decades whenever I have discussed Shiang's fascinating ideas in my university classes, my students have reacted precisely with a combination of glorious delight, shock, and delicious dismay.

I cannot overstate the importance of Shiang's work and its deep influence. It represents a paradigm shift in science not dissimilar to such epoch-changing concepts as the Theory of Relativity or the discovery that the earth is not flat.

A warning: This book is dangerous to some atheists and believers in randomness!

I never had much to say. Except to my students. Then David Shiang comes along. And David is the wildest thinker we have. And I just wrapped up in him.

– adapted from James Whitcomb Riley's
The Old Man and Jim

Len Klikunas
Department of Anthropology
Boise State University
Boise, Idaho
11 September 2007

1

Introduction

> We have become Antipodean in our scientific expectations. You believe in the God who plays dice, and I in complete law and order in a world which objectively exists, and which I, in a wildly speculative way, am trying to capture.... Even the great initial success of the quantum theory does not make me believe in the fundamental dice-game, although I am well aware that our younger colleagues interpret this as a consequence of senility. No doubt the day will come when we will see whose instinctive attitude was the correct one.
>
> *Albert Einstein, letter to Max Born, 1944*

This book is in many ways an expansion of ideas I first discussed in *On The Absence of Disorder in Nature*, published in 1979 on the occasion of the centennial of Albert Einstein's birth. At the time, I wanted to make my case in support of Einstein, who believed that we live in a world of complete law and order. As is well known, Einstein rejected the traditional view of quantum theory that nature is inherently probabilistic, and he lived the later years of his life more or less ostracized from the scientific community. Although Einstein is still celebrated as one of the greatest minds ever to have lived (he was called the "Person of the Century" by *Time Magazine* in December, 1999), his failure to embrace quantum theory is still met with scorn and derision on the part of many fellow scientists. They ask something like, How could Einstein have been so right about so many things and yet so wrong about quantum theory?

Few physicists think that Einstein was on the right path when he rejected quantum theory and insisted on law and order, but it turns out that he was correct when he said that "God does not play dice." Einstein, I maintain, will have the last laugh. The world is in order and never out of order. We will not be able to give Einstein everything he was looking for (his wish for unlimited prediction will remain unfulfilled, and the "spooky action at a distance" that he so dreaded has been confirmed), but we will give him enough.

I realize, of course, that concepts such as "right" and "correct" may have no place in science, which is often said to proceed asymptotically towards truth or reality. For example, Carl Sagan once maintained that "absolute certainty will always elude us." Such a sentiment may be true in science, but nevertheless we can know with 100% certainty that nature is governed by law and order, not chance and probability. If final answers can be achieved, as I affirm, we will have to go beyond science in order to achieve them.

We will not reject the scientific method, but we will see that it is quite limited concerning the kinds of questions it can answer. Science is not the only valuable line of inquiry into reality, despite what some scientists like to claim. And it turns out that science, despite its much discussed "self-correcting" mechanism, is based on various longstanding assumptions about nature that have no grounding in evidence and also happen to be completely wrong.

For example, "could the past have occurred otherwise?" is an important question that we all ask on occasion. The question cannot be answered through mathematics, logic, reason, or experiment, but it can be answered correctly nonetheless. And the correct (and final) answer has major implications for how we think about reality, time, the future, and human freedom.

Although we have an interest in knowing whether natural events such as hurricanes and quantum experiments could have occurred otherwise, what we really want to know is whether we ourselves could have acted differently in the past. Could we now be in a state other than the one we are actually in? Could we have chosen otherwise? Do we have genuine free will? The book's title—*God Does Not Play Dice: the Fulfillment of Einstein's Quest for Law and Order in Nature*—suggests that the answer to such questions is at hand.

Besides presenting both a qualitative and a quantitative solution to the free will problem (five words and three mathematical symbols, respectively), I will offer final answers to many other of our deepest mysteries, including the relationship between past, present, and future; the paradox of Schrödinger's cat; the nature vs. nurture controversy; and the question of design vs. chance in evolution. Departing from the conventional wisdom, I am going to challenge some of our most basic beliefs. (By the way, I do so somewhat reluctantly, as it would be much easier to maintain the status quo. I am by nature one who likes the path of least resistance.) I will force us to consider new, startling, and even unthinkable possibilities. However, the solutions I am going to put forth are entirely logical and, for the most part, quite simple.

A word of warning. Do not bother reading further if you do not have an open mind. I have on occasion "talked shop" with various physicists, and they tend fall into two types. The first type (a small minority, in my experience) quickly understands concepts such as CFD (Contra-Factual Definiteness) and the central role that it and similar hypotheses play in science. They are willing to question their assumptions and rethink longstanding positions. The second type (a vast majority, as far as I can tell) isn't really interested in anything that challenges their (supposedly secure) world view. They tend to get nervous and upset when I point out the existence of the hypothesis of equal *a priori* probabilities and the fact that although it has no experimental support, it is at the center of probability theory, statistical mechanics, the second law of thermodynamics, and many other sciences. They are uncomfortable with the notion that the hardest of the hard sciences might not be based on experimental evidence after all.

I once talked about the foundations of probability theory at a dinner where a physicist was present, and he nearly went berserk. His response went something like, "How could Las Vegas exist if what you are saying is true?" This poor soul didn't have a philosophical bone in his body (he had never heard of Karl Popper, for example) and was absolutely convinced that experiments had shown nature to be ruled by a fundamental chance. He didn't want to think clearly about the distinction between the individual event and the long run. In retrospect, it is plain that he had never been taught a thing about his own assumptions, and he wasn't going to start examining them now. He was perfectly happy to dismiss me as a "crank" (a favorite word among scientists used to disparage those who introduce new ideas that may appear to be unfounded and nonsensical), when all I was trying to do was show him the existence of a well-documented scientific hypothesis and what it meant for probability theory. He had never heard of the hypothesis, so in his mind it didn't exist. He was nearing retirement, and no one was going to rain on his "physics is as solid as a rock" parade.

I have often found scientists to be quite defensive when their theories and beliefs are questioned. They like to think that they deal in a world of solid facts and experimental observations. In my experience, they tend not to be very open-minded. They may talk about wanting to know how things really work and go on and on about how willing they are to entertain new ideas and change their minds, but they typically don't want to acknowledge their assumptions or beliefs, much less question them. Assumptions and beliefs are often hidden, and bringing them to light involves an introspection and self-examination seldom associated with scientists.

In *Quantum Reality: Beyond the New Physics*, Nick Herbert summed up the state of affairs in existence when he was in graduate school (and probably still largely true today). He wrote, "when I asked my teachers what quantum theory actually meant—that is what was the reality behind the mathematics—they told me that it was pointless for a physicist to ask questions about reality. Best to stick with the math and the experimental facts, they cautioned, and stop worrying about what was going on behind the scenes." If one isn't interested in what is really going on, one is to take refuge in quantification. This path has been advocated by many, and it seems to work. After all, quantum theory has been called the most successful theory in the history of science, never having failed an experimental test.

But as we shall see, a key contention of quantum theory is not subject to testing, despite what many scientists have claimed since the famous Solvay conference in Brussels in 1927. Scientists often use the term "mere philosophy" in a derogatory manner to refer to propositions that cannot be tested, but the notion that nature is inherently probabilistic falls squarely into this category. It turns out that there is no experimental evidence whatsoever to support such a viewpoint. One can certainly disagree with Einstein if one is so inclined, but it is false to say that experiments have shown him to be wrong. Despite what physicists such as Brian Greene and Stephen Hawking tell us, there is no evidence that God plays dice. Those who make such claims are confusing the kinds of predictions we can make about nature with the behavior of nature herself.

2

Beginning the Journey

Everything you know is wrong.

The Firesign Theatre

More than thirty-seven years ago, I began my undergraduate studies at MIT with thoughts of someday unlocking nature's secrets through the methods of science. Having been raised in an environment where science reigned supreme, I was eager to join a community that was unified in the pursuit of fundamental truths. Like many of my fellow students, I thought that a scientific career would be a sure route towards understanding the mysteries of the universe. I could think of nothing that was more important, and I relished the idea of contributing to an ever-growing body of knowledge.

As I progressed through my freshman year, however, I developed the uneasy feeling that something was missing amid all the math and formulas. It had seemed to me even before entering the citadel of science and technology that there were important areas of inquiry beyond the scope of the sciences, and I began to find my educational experience to be very limiting. As I got to know the atmosphere at MIT, where "hard" sciences like physics and chemistry were deified and everything else was considered second-rate at best, I became increasingly convinced that the scientific method was somehow inadequate as a means of fully understanding reality. Huston Smith, a former philosophy professor of mine, once told a story that will give you an idea of the atmosphere in existence at the time. One day during lunch at the MIT Faculty Club, Smith sat down next to a physicist. Deciding to engage in conversation, Smith asked, "What do you think of us humanists?" The physicist replied, "Think of you? We don't even bother to ignore you." I am not sure that much has changed since then.

I couldn't put my finger on any specific problem with my studies, but the more I learned about science and how it approached nature, the less I wanted to restrict myself to memorizing theories and manipulating mathematical equations. I don't mean to imply that my formal class work in calculus and physics wasn't valuable to me, for I found it extremely useful. I was tested and challenged in many ways, and MIT's reputation for teaching students how to use their analytical powers is, for the most part, well-deserved. (An MIT education is, by the way, often described as drinking water through a fire hose.)

On the other hand, I experienced an underlying discomfort because I didn't feel that I was allowed to do any original thinking in my science courses. I realized, of course, that students in the elementary stages of scientific study were supposed to learn an indisputable body of core knowledge and build from there. The foundations that we were being taught had stood the test of time, and supposedly there was nothing about them that needed questioning. Almost everything that we were expected to know could be gotten out of a textbook, often a very old one (or a new, more expensive one containing essentially the same ideas, but please don't get me started on the textbook market and why books that have little new material cost upwards of $100), and few fresh ideas were presented in classes.

As a result, many of us found that classroom affairs had a dull, repetitive feel to them. Nothing new seemed to be going on. Students could practically skip the lectures as long as they kept up with the text. The primary method of education focused on our obediently absorbing so-called "facts" and applying them mathematically under different circumstances. Problem solving was the order of the day. Our homework seemed to be an endless series of exercises dealing with artificial numerical problems. Correct answers to the puzzles that we tackled always existed, so there was little need for debate or discussion. We were simply mastering, in a step-by-step fashion, fundamental principles that had been established and accepted long ago. Challenging scientific authority was the last thing we freshmen were expected to do; our role at the bottom of the totem pole was to learn what we were taught with a minimum of commotion. Creative thinking would have to come later.

As a result of feeling as though I was being forced into a scientific straitjacket, I decided to carry out a search for knowledge as something of an extracurricular activity. I didn't have any specific goal in mind, but I knew that I wanted to enlarge my horizons and probe beyond the confines of science. I began reading dozens of books in a variety of other subjects, often spending hours on end at various bookstores in Cambridge and Boston. Among the disciplines I immersed myself in were philosophy, psychoanalysis, psychology, history, literature, and religion. While researching in these areas, I felt that I could formulate my own thoughts rather than simply learn a body of unquestioned knowledge. There was room for dialogue and inquiry of a kind that was wholly absent from my work in the sciences. My courses in math and physics were making me very good at coming up with solutions to all kinds of routine and insignificant problems, but I found such pursuits to be mundane and often tedious. I knew that I was gaining important grounding, but I really didn't care very much about calculating variable masses, inverse ratios, and angular velocities.

On the other hand, I recall the tremendous excitement of coming across Norman O. Brown's pioneering analysis in *Life Against Death* and Sigmund Freud's *The Interpretation of Dreams*. I remember examining Plato, Aristotle, Kant, Lao Tzu, and other philosophers who thought about notions of reality and the absolute. Colin Wilson's *The Outsider*, which I had read before coming to MIT, played an important role in my ongoing reflections of themes such as alienation and redemption. The work of T. S. Eliot also captivated me, especially the early poems that express a haunting vision of a human race at odds with itself.

As I continued to read practically anything I could get my hands on about the mind and various ways of approaching truth, I found myself making a kind of progress in my explorations. I slowly developed the feeling that I was onto something really important. I became increasingly confident that my search was leading somewhere. The more I worked at it over the months, the more it seemed that unrelated things fell into place. I had no inkling where this unfamiliar and at times intensely unsettling journey was going to end up, of course; I was simply taking in as much as I could from a wide range of sources. (I showed my reading list to Noam Chomsky, one of my professors, and he said something to the effect that it was a life's work.) Some of the material that I came upon would prove to be of no lasting value, but I wasn't about to rule out anything in advance. My general area of investigation was the unknown, and I was more or less prepared to follow any avenue that showed promise, no matter where it might lead.

About nine or ten months into my odyssey, I suddenly reached what I recognized at once as a major breakthrough. I experienced a momentous intellectual and emotional turning point that completely transformed my life. It was a shattering event in more ways than one. The many pieces of knowledge that I had been accumulating along the way were, figuratively speaking, abruptly turned upside down and inside out in a rush of definitive insight. All the assorted and often unrelated avenues that I had been following over my year-long search came together in a totally unanticipated fashion. I discovered, much to my surprise, that the world view that I had been embracing all along was full of illusion and error.

A new and unprecedented picture of reality forced itself on me, one that was all-encompassing, coherent, and perfectly logical. It was also preposterous, bizarre, outlandish, and seemingly impossible. I found it extraordinarily difficult to abandon a lifetime of beliefs in favor of conclusions reached almost overnight, but I knew that I had achieved a new and uncommon knowledge that was certain and final. It was as if I had found a higher level of consciousness or, to put it in another manner, a way out of the mental labyrinth of uncertainty and confusion. I was no longer in the position of grasping for answers; I knew that I had achieved them.

The experience of piercing what was in retrospect a veil of untruth involved a profound and terrifying exhilaration unlike any I had ever felt. Everything that I had taken for granted prior to that point needed to be reevaluated and put into another perspective, to say nothing of the scientific "facts" that I was in the process of learning. In no way could I have predicted my final destination in advance, as it involved a series of stunning cognitive and emotional reversals. At many points throughout the journey, I was "sure" that something was finally correct, but everything needed to be reconsidered at the ultimate breakthrough. A permanent sanctuary of peace and fulfillment awaited me, one free of philosophical doubt and uncertainty.

Given the peculiar nature of my findings, where many of our "common sense" perceptions about reality are in need of radical rethinking, the feelings of shock and astonishment I experienced were perhaps like those that Werner Heisenberg felt in the early days of quantum theory. In speaking of the tremendous difficulty he and other physicists had in accepting the results of new atomic experiments, he recalled a solitary walk he made in which he kept asking himself over and over if nature were really as "absurd" as seemed to be the case.

One purpose of *God Does Not Play Dice* is to convey something of the unique and penetrating experience that I went through many years ago. A similar journey is available to all of us, and I want to share the essence of my quest. My discussion may be of interest to those of you in the midst of your own search (and given the tremendous success of Google, it appears that everyone is searching for something); it may also engage those who are completely unaware that such experiences are even possible. Scientific types tend to fall into the latter category; they generally dismiss as inconsequential and trivial (a word often heard at MIT) anything that doesn't fit into their preconceived notions of how nature works. I must confess to being amazed at how narrow-minded some of them are when they insist that science (materialism) is the only road to reality and that other means of acquiring knowledge are irrelevant.

A second and equally significant purpose of this book is to discuss the various findings about nature and reality that result from the experience. The notion of ascertaining deep truths through the mind may sound far-fetched, but it is the revolutionary nature of the experience that makes this kind of knowledge possible. The human mind has the capacity to understand fundamental reality, and as we explore questions about what kind of knowledge we can and cannot possess, I will argue that many of our deepest insights are a result of experience, not mathematics, measurement, or experiment. There is, of course, a long list of important thinkers who trusted in the primacy of the mind, including Einstein, Max Planck, and Erwin Schrödinger.

But I do not simply want to criticize the scientific method and show its limitations; I intend to go much further and challenge the fundamentals of science itself. It may seem heretical to question principles that go back more than three hundred years, but I am going to suggest, quite seriously, that the modern scientific framework has been constructed upon a partially-faulty foundation. Perhaps it is presumptuous to challenge ideas that have been spoon-fed to generations of students, but I will maintain that scientists have made many false assumptions about nature and that as a consequence, the laws of physics are full of unreality. The so-called "success" of quantum theory has given us many useful inventions, but it does not imply an actual understanding of nature.

Many of you are undoubtedly accustomed to the ubiquitous ideas put forth by scientists and mathematicians that probability governs the world and that nature inexorably progresses towards increasing randomness and disorder. The second law of thermodynamics may have found its way into all of the sciences and even many aspects of popular culture, but I will allege that the mathematical abstraction that the scientists call "disorder" is a figment of their imagination. To be more precise, I will claim that "disorder" has a constant value of zero (0). Since we are dealing with ideas that are deeply ingrained into our thinking (often perhaps without our knowledge), deconstructing them in order to show their true origins will take some doing.

In *Descartes' Dream: The World According to Mathematics*, Philip Davis and Reuben Hersh write that "the stochastization of the world (forgive this tongue-twister) means the adoption of a point of view wherein randomness or chance or probability is perceived as a real, objective and fundamental aspect of the world." It turns out that randomness, chance, and probability are not real, despite what Davis and Hersh claim. I will show in no uncertain terms that the scientists start with a very questionable assumption about nature that is by its very definition unscientific. In no way can their fundamental starting point be considered "objective," although it may well be logical and reasonable. Once we have leveled the playing field, so to speak, by showing that the scientist's world is based on presupposition rather than evidence, we will be in a better position to judge how well scientists follow their own rules. Ironically, we will find that the "hardest" of the "hard" sciences contains nothing but a metaphysical assertion at its core.

In addition to calling into question various claims of physicists and mathematicians, we must also challenge Darwinian theory and the views of evolutionary biologists. If you follow in the footsteps of Darwinian defenders such as Richard Dawkins and Jared Diamond, you may be entirely comfortable with the idea that our very existence is due to assorted happy and fortuitous accidents of cosmic architecture as well as wildly improbable evolutionary happenings on earth. Staples of science such as chance, accident, serendipity, odds, and blind luck may form integral parts of your belief system, but I will contend that they have exactly and precisely nothing (zero) to do with reality.

Ernst Mayr tells us in *What Evolution Is* that "there is indeed a great deal of randomness ('chance') in evolution" and that "chance reigns supreme" at the first step of the selection process. Unfortunately for Mayr and his colleagues (but perhaps not the rest of us), this view of evolution happens to be wrong. We can know with certainty that variation (and everything else that follows) occurs by design, not by chance.

Whatever your viewpoint, I am going to invite you to think again. Much of our so-called "conventional wisdom" will be set on its head in the world that you are about to enter, where the unexpected, the bizarre, and the outlandish will be in abundance. If I can summon any defense for my contrarian outlook, it is that I did not ask to be put in the position of challenging an entire tradition. I may be a revolutionary, but I am a reluctant one. I have no desire to offend anyone (or everyone)—I am simply interested in understanding the world as it really is, no matter where things may lead. If it turns out that the laws of nature and the laws of physics have no relation to one another at times, then the discrepancies should be illuminated and brought into the open. If it turns out that the Darwinian theorists have led us down a false path, I should think they would want to know. In any case, much of our understanding of nature needs to be fundamentally revised, and what may appear to be crazy needs to be considered seriously.

You are of course entirely welcome to disagree with the radical solutions that I propose to various problems, but I think you will find that many of my arguments, such as those dealing with hidden beliefs and assumptions in physics and evolutionary biology, are unassailable. I will examine the foundations underlying the rationalist world view and will point out that much of their presumed solidity is about as solid as the space between galaxies. Scientists may have convinced themselves and many of the rest of us that they are impartial and dispassionate observers in search of nature's deepest secrets (and to some extent this is the case), but their theoretical model has been biased from the start. (The fact that the model is incorrect is another issue altogether. It should be obvious that bias does *not necessarily* mean incorrect.)

For the most part, scientists are wholly unaware that the foundation of their world view rests upon assumption and conjecture, not experimental result. Far from approaching nature with a neutral and unprejudiced attitude, scientists begin with largely-unquestioned presumptions that completely color their perspective. Let me use the following metaphor: the scientist approaches nature wearing "chance colored" eyeglasses prior to conducting any experiments whatsoever. These glasses define how the scientist looks at the world, but the scientist does not realize that he or she has them on. We will see that scientific claims of "objectivity" when it comes to characterizing nature are at times little more than a fabrication, even in the context of classical physics. As is well known, quantum mechanics has thrown the concept of objectivity at the atomic level into serious doubt.

In my four years at MIT, not once did we consider the underlying ground rules that form the basis of the scientific world view. I am not even sure that most of my professors were aware of their preconceived notions, as these notions are part and parcel of the traditional, unspoken outlook. Rigorous impartiality on their part was taken for granted, and assumptions dating back centuries were left unexamined and untaught.

It is curious, but some scientific biases have a distinctly Western quality, and their origins can be traced back to the ancient Greeks. For example, Chaos, the Goddess of Emptiness and Confusion, arises from Greek mythology; no self-respecting Taoist would embrace such a concept. Democritus is credited with introducing the interplay between chance and necessity, a concept that survives largely unchanged to this day. And Aristotle discussed "potentia," or potential, paving the way for the still-continuing debate about the possible versus the actual.

Despite what some scientists may claim, the scientific method is not even capable of dealing with many fundamental issues that must be addressed in any comprehensive understanding of nature.

As you read through this book, I expect that you will come away with the realization that a lot, an awful lot, of what passes for "fact" in the world view of the scientist is nothing more than unsupported and unsupportable assumption.

Given the nature of recent advances in physics, the kind of "craziness" that I am proposing is perhaps exactly what physicists, following the line of argument of Niels Bohr, should be expecting. Shortly before his death in 1962, Bohr is reported to have said the following about a new proposal: "We seem to all agree that the theory is crazy. The question is, is it crazy enough to have a chance of being right? My own feeling is that it is not quite *that* crazy" (emphasis his). There is much more than an element of humor here, as anyone familiar with the history of science knows. Regarding my own views, I contend that what I have to say is "crazy enough" to be right and also provides exactly the kind of weird simplicity that might have been anticipated. Although nature is under no obligation to have in place laws that we should find acceptable, I think you will agree that in the world view that I set forth, she can be viewed as entirely reasonable and extraordinarily coherent. We may not like the picture I am painting, but we cannot say that it is illogical or irrational.

3

A Certain World Beyond Science

> A treasure stumbled upon, suddenly; not gradually accumulated, by adding one to one. The accumulation of learning, "adding to the sum-total of human knowledge"; lay that burden down, that baggage, that impediment. Take nothing for your journey; travel light.
>
> <div align="right">Norman O. Brown, <i>Love's Body</i></div>

Many different descriptions have been used to characterize the mental voyage that I went through, including "transcendence" and "the highest state of consciousness." Such terms might be appropriate in some settings, but I prefer to approach this formidable yet delicate subject with a vocabulary that is as free as possible from preconceived notions. The above terms have their value, but such descriptions are normally tainted with misleading, unwarranted, and negative connotations. For example, the concept of "transcendence" in Eastern philosophy, which is used to symbolize the realization of what is referred to as "ultimate reality," often mistakenly carries with it ideas of irrationality and illogic. We are told that this experience and the insights gained are beyond the scope of rational investigation and verbal communication. It is presumed that such a phenomenon, if it can be talked about at all, is closed to analytical scrutiny.

Judging from typical writings on the subject, which are often full of seeming nonsense and apparent contradictions (e.g., what is the sound of one hand clapping?), it is easy to see why those who cherish rationality and logic (and I include myself among them) would want to have nothing to do with this method of perception. After all, those of us who instinctively feel that nature is governed by reasonable laws can and probably should dismiss whatever is said to be illogical to begin with. Let me say at the outset that higher consciousness, properly understood, has nothing to do with illogic, irrationality, anti-science, or anti-reason. It is, rather, a knowledge beyond that which can be gained by measurement, reason, and experiment. It is knowledge beyond the scientific method.

If there are potential difficulties in using such designations as "transcendence," perhaps we should seek a different kind of description. "Mind of God" is certainly useful and has been used by any number of thinkers such as Einstein and Stephen W. Hawking. Einstein wanted "to know how God created this world." Hawking suggested that "the ultimate triumph of human reason" is to "know the mind of God." However, God means different things to different people. Einstein and perhaps Hawking see the term as a representation of the universe's order and harmony. Others view God as a benevolent being who cares about humans and actively intervenes in their daily lives as a direct result of prayer and devotion. (This latter God seems to be what Steven Weinberg so steadfastly and vehemently rejects. As we shall see, we have no use for this type of supernatural God either.)

Due to the lack of agreement about the meaning of "God," let us use the phrase "gold mine of consciousness." It appropriately combines images of precious and dazzling treasure with the active workings of the mind. In addition, there is apt to be a minimal amount of unwarranted philosophical baggage associated with this term, unlike with the word "God." There is nothing necessarily irrational about the idea of a state of consciousness in which one gains significant insights into nature and reality. (If you equate "spiritual," "religious," or "mystical" encounters with nonsense or with "temporal lobe seizures or some other aberration in brain physiology," as Michael Shermer does in *How We Believe: The Search for God in an Age of Science,* please set your views aside for the time being. Try not to bring too many preconceived notions with you on our journey, as they are likely to get in the way. You can always go back to your original outlook later if you so desire.) Finally, "gold mine of consciousness" is not limited to a specific culture or historical period, and it does not imply an explicit philosophical or religious tradition. I will maintain later that the question of God is ultimately one of knowledge vs. ignorance, not faith vs. reason or belief vs. denial, but it is, I think, better not to begin with the concept of God as a prerequisite.

I can imagine that my use of the terms "transcendence" and "gold mine of consciousness" has some of you wondering where all of this is heading. Am I going to abandon priceless rationality and logic in favor of some sort of mystical mumbo jumbo of the sort found in what we often call "occult" books? Believe me, I too have little use for ESP, channeling, UFOs, levitation, pyramids, crystals, or astrology. Are the solutions to riddles that I promise simply the obtuse and nonsensical ravings of a lunatic, or even worse? As one schooled within the scientific tradition, where skepticism rightfully abounds, I empathize with any feelings of doubt that you may have at this point. In fact, I wholeheartedly welcome your reservations. But please don't develop a case of premature closing of the mind. (If you can cite evidence where I say that none exists, I would be grateful. I don't think you can do it, however.)

While I have great admiration for the scientific method (logic, reason, hypothesis formulation, experiment, etc.) and want you to maintain your critical faculties to the fullest here, let me suggest that your possible lack of familiarity with my subject matter in no way diminishes it. Remember the words of philosopher and logician Alfred Jules Ayer, who once said that "we can hardly maintain *a priori* that there are no ways of discovering true propositions except those which we ourselves employ." My suggestion of a method that you don't know much about may be surprising, and the conclusions I put forward may be extraordinary, but this does not weaken my case in the least.

To some of you, the words "gold mine of consciousness" may conjure up visions of a blissful, transcendental state involving a feeling of unification with the cosmos. All of us are familiar with wise-looking gurus dressed in flowing robes who promise us inner peace and eternal "I see the light" happiness. And evangelists preach that "the kingdom of God" is within us. Judging from the abundance of spiritual activity around us, the selling of salvation is a major growth industry. There are good reasons for this phenomenon, even though some of those who would save us may not have the best of intentions. In any case, our subject is not really new; it has long been explored by many thinkers from different cultures and backgrounds. Older texts such as the *Upanishads*, the *Tao Te Ching*, or the *Bible* as well as relatively contemporary discussions by insightful authors such as Richard M. Bucke, Alan Watts, Jean Houston, Fritjof Capra, John White, and Charles Tart are quite pertinent to our discussion.

Although there is much of value in the diverse literature, I think that it is better to proceed without stopping for an examination. On one hand, I do not want to duplicate what others have said. There are countless descriptions of euphoria and universal harmony that may be of some use in understanding such a state, and I see no need to cover the same ground. On the other hand, many authors who write about the "gold mine," perhaps using a different terminology, tend to suggest that it is "ineffable" or "non-intellectual." They almost admit a kind of defeat concerning intellectual discourse before they begin. The words "irrational" and "illogical" are often found in this context, closing off all attempts at intelligent discussion. (No wonder scientists and many others have a hard time dealing with the subject.)

If we were to rely on the typical work about higher consciousness as a reference, it is likely that we would spend far too much time examining points of debate. Some writings are so cryptic that few can agree on just what they mean. In other cases, the points of view expressed are in direct opposition to my own, particularly regarding what can be communicated about the experience. We may therefore be better off taking an approach that is somewhat different from the ones already mentioned. I propose that we use the mythological perspective, which can serve as a useful guide.

Let us look at the "gold mine" from two viewpoints. The first has to do with the process of psychological transformation that one undergoes as one experiences the "gold mine." The second focuses on the contents one discovers inside the "gold mine." The mythological journey of the hero will serve as a useful aid in probing the central features of the quest. Considering that many different avenues can lead to the "gold mine," the heroic archetype provides us with important signposts in making sense of a most bewildering ordeal.

The hero, the obstacle that must be overcome (often represented by a dragon), and the treasure are the three core elements in the struggle, and they are at the center of a myriad of seemingly unrelated myths. Joseph Campbell, Mircea Eliade, and others have done an admirable job of noting the common themes underlying hero myths throughout history, and their work can help us delve more deeply into the significance behind the symbols.

In its simplest form, the hero myth involves someone of unusual capacity who undergoes a most difficult journey, overcoming many obstacles on the way to attaining great treasure. Jason, Heracles, Perseus, and Odysseus are notable examples from ancient Greece. Oedipus is one of the most celebrated heroes in all mythology, as he slays the Sphinx and is rewarded with the hand of the Queen. In this case, the archetypal obstacle (dragon) takes on the form of the menacing Sphinx, and the treasure is the Queen. (Oedipus, of course, does not know at first that he married his own mother, leading to all sorts of problems down the road. Did someone say tragedy? A minor digression: one significant aspect of Freud's "discovery" is that he formulated his incestpatricide theory *before* consulting the Oedipus legend, as his own letters show. Like many of us, scientists and non-scientists alike, he went to the evidence in search of confirmation, not illumination. And he found what he was looking for. I examine in detail Freud's formulation of the Oedipus complex in *Weird Scenes Inside the Gold Mine*.)

The dragon can have any number of external faces, but for our purposes its importance lies in the fact that it symbolizes psychic obstacles that must be overcome in order to achieve psychological transcendence and liberation. The physical represents the mental. Obstacles can be as diverse as false temptations, monsters, blind alleys, and illusions. By slaying the dragon, the hero achieves breakthrough and gains the treasure. The reward can take many forms: a maiden, gold, diamonds, power and glory, the elixir of life, immortality, etc.

Our second perspective deals with the bizarre contents of the "gold mine" itself. Using the mythological journey of the hero as a metaphor, the end point is of immeasurable importance. It can be described as the attainment of higher consciousness and psychological fulfillment—a treasure trove of rich insights into the nature of reality. Here, deep mysteries such as the relationship between past, present, and future are laid bare and solved. In addition, one realizes that we are here as a result of a creator or designer and that design is a primary feature of the world. The notion that things happen by chance and accident is seen to be a false assumption based on ignorance.

Certainty and knowledge replace speculation and supposition, and an unprecedented level of understanding is achieved. Unlikely as it may seem, there is a vast amount of illuminating content that comes from the experience. The expedition we are about to embark on should help you if you have an interest. The more one considers the knowledge that is gained upon reaching the "gold mine," the easier it will be to make the experience one's own.

You may have noticed my use of the word "design" above. If you accept such a notion (if only for the time being), you are, of course, welcome to call the designer "intelligent," "dumb," or anything else, depending on your point of view. Stephen Jay Gould points to "odd arrangements and funny solutions" as "paths that a sensible God would never tread." Shermer tells us that "the eye has evolved independently a dozen different times through its own unique pathways, so this alone tells us that no creator had a single, master plan." Such thinking is a staple of many in the anti-God community. However, who is to say that God should be bound by Gould's definition of "sensible"? Gould and Shermer expect a designer to conform to their ideal of a perfect (or at least intelligent) engineer, and when they find that some aspects of life appear to have been put together using tape and glue as core components, they insist that there is no God. No intelligent designer would be so stupid, so dumb, so lacking, they claim. They see "mistakes" in nature as evidence for natural selection. Such thinking is perfectly logical, but it has nothing to do with reality. What if God's master plan includes what some perceive to be bad design and poor planning? What if God actually intended multiple pathways for the eye? We will explore these ideas in more detail later.

4

The Solution to the Free Will Problem

I am compelled to act as if free will existed, because if I wish to live in a civilized society I must act responsibly.

Albert Einstein

Imagine, for a moment, that we live in a world where humans have no power to affect the future. A world where all decisions and all outcomes, no matter how momentous or how trivial, are already determined (or predetermined). A world where the history of the future is already written, so to speak. Such a world may sound frightening or even dreadful. After all, it would mean that we were not really "responsible" for our past glories. We would be unable to claim that our celebrated accomplishments were completely of our own doing. Frank Sinatra's "I did it my way" would take on a whole new meaning. And such an apparent lack of freedom would mean an inability to control our own destinies. In no way could we call ourselves masters of the universe (a favorite term used by Wall Street traders to describe themselves), free to shape the world as we see fit. One of our most cherished desires, total freedom of the will, would be nothing more than a pipe dream. Some might say that in such a world we would be like puppets, simply going through the motions of a prepared script. Humans would be reduced to mere actors on a stage, having no original lines of their own. To many of us, such a world is unthinkable and deplorable. *Nevertheless, I am going to insist that this is the world we have always been living in.*

This assertion may appear to be very strange indeed and therefore difficult to take seriously. I mentioned earlier that a key mystery related to the nature of time—the relationship between past, present, and future—is solved in experiencing the "gold mine." We all have an easy time admitting that we cannot affect the past; I am going to insist that there is a basic symmetry here and that the solution to the free will problem can be summed up in five simple words as follows:

> You cannot affect the future.

These five words that can literally change your life forever. Another way of putting it is that the world has been designed in such a way that you are completely powerless when it comes to controlling what will take place in the future. In other words, the only future outcome that *can* occur is the outcome that *will* occur.

Few of us will find it easy to accept this notion—most of us want to be in full control of our own destinies. We want to think of "endless possibilities" in a world where we are completely free to do as we please. "Control your own destiny" is an often-expressed mantra, especially in western society. We recognize that some phenomena such as the weather and the next spin of the roulette wheel may be beyond our control, and we think that chance, randomness, and luck may take us in unexpected and unpredictable directions, but we are sure that much is within our sphere of influence. (Much of science is about expanding our so-called control of nature.)

In *Darwin's Dangerous Idea: Evolution and the Meanings of Life*, Daniel C. Dennett discusses the free will problem and its bearing upon evolution. Using the term "actualism" (the idea that only the actual is possible) introduced by Ayer (whom we met earlier), Dennett calls it a "dread hypothesis" and an "old nemesis" and claims to offer good reasons for "dismissing" it. Dennett wants his "elbow room"; he wants to be in control. (*Elbow Room: The Varieties of Free Will Worth Wanting* is one of Dennett's other books.) He writes, "I am prepared to assume that actualism is false (and that this assumption is independent of the determinism/indeterminism question), even if I can't claim to prove it, if only because the alternative would be to give up and go play golf or something."

I suggest that Dennett get his hands on a set of indestructible golf clubs, although I do not share his pessimism or cynicism about such a world. Dennett's "give up and go play golf" reaction to the idea that the actual is the only possibility or that we have no free will is not at all uncommon. Such a simplistic viewpoint is also expressed in the romantic comedy *Serendipity*, a delightful movie about two people (John Cusack and Kate Beckinsdale) who meet "by chance" and whose paths cross again years later partly due to "fate" or "destiny." The character Eve (played by Molly Shannon) asks why we should "get out of bed in the morning" if we are not really free, which is essentially the same as Dennett's resignation. We will examine Dennett's position in more depth in a subsequent chapter, but I find it surprisingly shallow for a cognitive philosopher of his stature, position, and credentials. The absence of free will has many positive attributes, but Dennett appears not to have thought about them. Above all, he wants to be in control. If this power is taken away, he believes that all is lost.

Notice, by the way, the simplicity, economy, elegance, symmetry, comprehensiveness, and lack of ambiguity regarding the solution to the free will problem that I am proposing (which happens to be the correct and only solution). Such aesthetic characteristics are, of course, not a defense of the solution but an added· bonus. (The solution is correct and does not need "defending" in the traditional scientific sense. On the other hand, we would not expect the solution to be clumsy, awkward, and inelegant.)

> You cannot affect the past;
> You cannot affect the future.

Things could hardly be more simple, although living it is another matter. (It's like what they say about Texas Hold 'Em Poker—it takes a minute to learn and a lifetime to master.)

Many of you may think that it is impossible that the world could be constructed in such a way, for it would seem to indicate that what we call free will is merely an illusion. I like freedom and elbow room as much as the next person, but I have bad news for all of you who like to think that you are the master of your own destiny. You are in store for a massive and overwhelming deflation of the ego. As The Firesign Theatre put it, "We're all a lot shorter than we used to be." There is no such thing as free will. However, I will argue that we have all the free will that we need (which is zero). We just may not have as much as we want.

Max Born, writing to Einstein, called the idea of a world without free will "quite abhorrent" and also (incorrectly) thought that it undermined the concept of human morality. Robert M. Augros and George N. Stanciu maintain in *The New Story of Science: Mind and the Universe* that "a denial of free will would render the whole of science absurd." I see their point, but I think they profoundly misunderstand what the lack of free will really means. It does not mean being manipulated against one's will. It does not mean feeling like you are a puppet. (As an aside, Einstein more than once explicitly said he did not believe in free will. He once told an interviewer, "I am a determinist. I do not believe in free will," and he repeatedly debated this point, along with morality and responsibility, with Max and Hedwig Born.) There are many advantages to not being in control, especially when it comes to concepts such as regret, remorse, and guilt. Always wondering "could I have done otherwise?" is a sure route to unending anxiety and uncertainty. Examples abound of people who have gone to their graves wondering "what if"?

One element of the "gold mine of consciousness" experience involves using our memory in a very unusual way. Specifically, one *remembers* that the past has occurred in the only possible manner. One confirms the notion of actualism, the view that the actual, the real, is the only possibility. Dennett poses the question, "Could anything happen other than what actually happens?" The answer is a resounding and unequivocal "No." One finds that nothing could have taken place in any way other than the way it actually happened. "Could have," with all its ghosts and shadows, is reduced to the actual. No more uncertainty and speculation about "what might have been."

The shocking realization that the past could not have taken place otherwise is made by consciously exploring what has been called the "collective unconscious" or "universal mind." There is a wealth of knowledge in the unconscious that extends far beyond the life of the individual—some of this knowledge has to do with the past. Norman O. Brown provides a useful point of view in the following passage from *Love's Body*:

> The unconscious, then, is not a closet full of skeletons in the private house of the individual mind; it is not even, finally, a cave full of dreams and ghosts in which, like Plato's prisoners, most of us spend most of our lives—
>
> The unconscious is rather that immortal sea which brought us hither; intimations of which are given in moments of 'oceanic feeling'; one sea of energy or instinct; embracing all mankind, without distinction of race, language, or culture; and embracing all the generations of Adam, past, present, and future, in one phylogenetic heritage..."

Reaching the "gold mine" requires exploring the heritage of the unconscious and bringing it to consciousness, and it calls for the use of memory. This isn't the ordinary use of memory, of course, but it is the use of memory nonetheless. And the memory that the past could not have occurred otherwise pertains to our own actions as well as all other phenomena throughout the universe.

The actual (real) situation we find ourselves in today, whether arrived at in a straightforward manner or through what may be (erroneously) seen as a series of "chance encounters," "improbable accidents," "amazing coincidences," "astronomical odds," and "fortuitous events," was bound to occur. The present was wholly unavoidable and unpreventable. Physicists and evolutionary biologists alike are constantly telling us how lucky we are to be alive and how we have overcome huge and overwhelming odds, but luck has nothing to do with it. The mathematically wild improbabilities they calculate are based on a fundamental misreading of reality, little more.

My view of past, present, and future is, of course, in stark contrast to the way we normally think. Most of us presume that things could have turned out quite differently. Dennett, citing Hume, defines "elbow room" as "the looseness that prevents the possible from shrinking tightly around the actual." Most of us, including Dennett, want that looseness to be quite roomy so that we can play out our endless "alternative scenario" thinking and feel that we are in control. The words "what if" play a very important part in how we think about the past and the future. Of course we could have acted differently, we say. Of course we had the power to do otherwise. We think we could have done something we did not do, and we think we could have not done something that we did do. (The uncertainty works both ways and covers the entire landscape.)

One need only look at some of the latest books and periodicals to see how speculations about "what might have been" preoccupy us. What if your parents hadn't met? What if President Kennedy hadn't gone to Dallas? What if Hitler hadn't been born? What if Bill Buckner had caught the ground ball that ended up going through his legs in the 1986 World Series? What if Hurricane

Katrina hadn't hit New Orleans? What if the earth were a little farther from the sun?

Although such questions may be of interest, the answers are largely remote and abstract. Of more relevance are speculations about our own personal situations. We think that because we make choices, we could have done something different that would now put us in another reality. What if I had gone to a different college? What if I had chosen the other profession I was considering? What if I had married someone else? What if I had bought the winning lottery ticket? What if I had hadn't missed the flight that crashed? I suggest that all of us have wondered about questions such as these at one time or another. There is a lot of fuzziness in how we view the present, as the notion that we could be in a different reality if we had acted otherwise is very much a part of our ingrained thinking.

The fuzziness extends to events that we think are outside of our control. We are quite sure that many external events could have occurred otherwise. You are reading my words at this very moment, but according to Dawkins, you are doing so against "astronomical odds" and are "unthinkably lucky." Dawkins writes in *Unweaving the Rainbow: Science, Delusion, and the Appetite for Wonder,* "the lottery starts before we are conceived. Your parents had to meet, and the conception of each was as improbable as your own. And so on back, through your four grandparents and eight great grandparents, back to where it doesn't bear thinking about.... the humblest medieval peasant had only to sneeze in order to affect something which changed something else which, after a long chain reaction, led to the consequence that one of your would-be ancestors failed to be your ancestor and became someone else's instead.... the thread of historical events by which our existence hangs is wincingly tenuous." (Please keep this phrase in mind, as I am going to discuss it in more detail later.)

Dawkins tells us that even though we are alive at this moment in history, the odds are against it. There is (quite obviously) no evidence that things could have turned out otherwise, but Dawkins is secure in his belief that "the potential people who could have been here in my place but who will in fact never see the light of day outnumber the sand grains of Arabia." We will examine Dawkins' wholly speculative viewpoint when we utilize the tools contained in Sagan's baloney detection kit. For the time being, ask yourself whether "potential people" lies in the realm of fact, fiction, or science fiction.

Others such as Gould and Murray Gell-Mann join Dawkins in thinking that whimsical "twists of fate" or "accidents of life" could have led to different circumstances. Contingency thinking is said to be a hot area of inquiry among historians and evolutionary biologists alike (what would have happened if?) and has been explored in some detail by Dennett, Dawkins, and Diamond, among others. Unfortunately, this line of reasoning, although occasionally interesting and perhaps even at times fascinating, is based on a false assumption and has very little to do with reality. The tree of life could not have taken any unreal detours. The real is the only possibility. History could not have been otherwise. No wonder Dennett tells us that actualism would make a "mockery of the Darwinian claim" and that Darwinism would be "defunct." These are, of course, very serious (and accurate) charges, and we will explore them in Chapter 8. It turns out that Dennett is an unwitting ally of anti-Darwinians such as myself, as he has clearly articulated the disastrous consequences for evolutionary biology if the real is the only possibility.

The enormous, incalculable, and often disquieting uncertainty that accompanies thoughts such as "if only I had done things differently" or "it could be otherwise" takes on a whole new dimension in the predetermined world that I claim we live in. Much of our attention is spent second-guessing the past, but we can know with absolute certainty that nothing could have taken place in a different manner. This significant and even stunning recognition eliminates a lot of the anxiety and confusion that occupy our lives.

By one estimate, more than 30% of the time we spend worrying concerns things that have happened and can't be changed. A simple alteration in the way we think can lead to the virtual elimination of this entire type of worrying. No longer do we need to be haunted by the omnipresent ghosts of "what might have been," "why did I do that" or "why didn't I do that," as such speculative "possibilities" fade away completely. The skeletons in our closet that continue to cause us anguish, if not misery, can be placed in their proper perspective. This does not erase the fact of our having done or not having done something, but it does offer us a genuine freedom (of a particular kind) from the past. I may not have bought Microsoft stock when it went public, but I can safely say that I could not have done otherwise. Why should I regret my so-called "missed opportunity"?

Strange as it may seem, memory can also give us knowledge about the future. We do not have foreknowledge of exactly what *will* happen (a key goal of science), but we do know that all future events are predetermined and unchangeable. One logical consequence of experiencing that the past could not have been different is realizing that the future has already been written, so to speak. This should be intuitively obvious. Think about it in the following way: the events of the past could not have gone any differently, despite all the informed "choices" or unplanned "accidents" that "affected" the outcome. This statement is timeless, or independent of time. Since we are able to say at a future date (for example, tomorrow) that the past was predetermined, then all the events between now and that future date have already been predetermined.

This realization leads to an understanding about the future that results in a kind of freedom and inner peace. Our future may be "before us" and not yet traveled, but the road has already been designed. In *Little Gidding*, Eliot tells us that this use of memory gives us liberation from the future as well as the past.

> This is the use of memory:
> For liberation—not less of love but expanding
> Of love beyond desire, and so liberation
> From the future as well as the past.

We all agree that we cannot affect the past; through the use of memory, we make the stunning discovery that we cannot affect the future either. If there is any term that sums up the philosophical component of the "gold mine," it is "predeterminism." I hesitate to use the word "determinism" because of the heavy baggage that this word carries, as it is often equated with "predictability." We will see that a predetermined world does not in any way mean a world that should be predictable to physicists or anyone else.

We are not in control, we have never been in control, and we will never be in control. So much for the scientist's desire to control nature, an aim that was doomed from the start (and highly dubious, in my opinion). The interplay between humans and nature is far more subtle than most scientists realize. No doubt most of us "feel that we can do something to affect the future," as Richard Feynman notes in *The Character of Physical Law*, but complete symmetry exists here, and we cannot affect the future any more than we can affect the past.

The answer to the "free will problem," again, is that there is no such thing as "free will." This solution, as distasteful as it may seem to some of you, is not without its merits. It happens to solve the longstanding philosophical dilemma regarding the "law of excluded middle," which, when taken to its logical conclusion, indicates that we have no free will. In order to continue to believe in free will, philosophers decided that they would "reject the common practice of conceiving truth as a timeless characteristic of propositions." By doing so, "only then can our belief that it is within our power to affect the future be sustained," Bernard Berofsky writes in *Free Will and Determinism*. As you can see, the belief in free will is a very powerful and dominant one. Philosophers would rather cling to freedom at the expense of their own logic. They would rather change their longstanding definition of truth in order to preserve free will.

In addition, according to John Bell (of Bell's Theorem), who used the term "super-deterministic," the lack of free will solves a fundamental mystery posed by quantum experiments. The physicist assumes that he or she can choose which experiment to perform, but if this is not the case and there is no real choice, then "the difficulty which this experimental result creates disappears." In response to an interviewer's question "Free will is an illusion – that gets us out of the crisis, does it?," Bell answers "That's correct." This elegant and uncomplicated solution to many mysteries of the quantum world is of the utmost importance, but it is hardly taken seriously by the vast majority of physicists. They would rather reject Bell's "super-deterministic" explanation (despite its beauty, symmetry, and simplicity) and keep the crisis of quantum theory alive and well, as they want to preserve free will. They would rather come up with clumsy, complicated, elaborate, twisted, and convoluted explanations that, among other things, allow them to continue believing in their own ability to control. Just the other day, I read a long-winded article that examined eight different solutions to quantum riddles. Not one of these solutions mentioned the simple super-determinism put forth by Bell. The multiverse and other inelegant explanations were, however, featured prominently, and they all allowed the physicist to believe in free will. (See the Notes section for an illuminating exchange on free will between Bell and P.C.W. Davies.)

We may be able to think of alternative paths of action and make choices based upon a rational (or irrational) decision-making apparatus, but this doesn't mean that we can affect the future. Feynman tells us that "remorse and regret and hope and so forth are all words which distinguish perfectly obviously the past and the future," as if to suggest that all is self-evident and that nothing more needs to be said. There is certainly a big difference between remorse, which looks towards the past, and hope, which looks towards the future. These concepts, however, take on a very different meaning when we make the realization that the past and the future contain fundamental similarities and that we cannot influence either. Remorse becomes completely transformed when one realizes that one could not have done otherwise. (Psychologist and leading self-help expert Hamilton Beazley, Ph.D., has written a book called *No Regrets: A Ten Step Program for Living in the Present and Leaving the Past Behind*. Although the book is interesting and occasionally useful, the problem of regret cannot be solved until and unless one realizes that one could not have acted otherwise. Beazley's solutions are ultimately only cosmetic and do not address the source of the problem. Only one step is really necessary to let go of regret, which is changing how one looks at the past.)

In *Order Out of Chaos: Man's New Dialogue with Nature*, Ilya Prigogine and Isabelle Stengers make a statement that goes to the heart of the matter: "Even the scientist who is convinced of the validity of deterministic descriptions would probably hesitate to imply that at the very moment of the Big Bang, the moment of creation of the universe as we know it, the date of the publication of this book was already inscribed in the laws of nature." Actually, the "Mind of God" or "gold mine" tells us precisely what Prigogine and Stengers would deny. We can know with certainty that from the very beginning, events began to unfold according to a predetermined time and pattern, and the publication date of their book as well as the one you are now reading was already cast in stone, so to speak. This does not mean that Newton's mechanistic view of the world was correct in all its aspects; the belief in total prediction (in theory) was a futile one all along. (We will look more closely at the downfall of classical physics later.) Nevertheless, the idea of a universe unfolding according to a predetermined plan still has its usefulness.

At first glance, it may appear that human freedom has no place in a predetermined world. Nothing could be further from the truth, although we must define precisely what we mean by "freedom." It can be said that all future events are completely predetermined, from the place where the next electron will appear in an atomic experiment to the winners of all future Olympic Games, but humans still make choices and act upon them. We feel free and go about acting as though there are no artificial constraints that limit our activities, but we are acting in the only manner possible.

We have the feeling of being free but we are not free.

I myself find this an ingenious state of affairs (brilliantly simple, one might say), but you can find it objectionable if you so desire. However, if given the choice, I think most of us would rather feel free than be free.

Perhaps there is a paradox here. (No one ever said that nature was straightforward.) We are reminded of Einstein's "*Raffiniert ist der Herr Gott, aber boshaft ist er nicht*," which has been translated in various ways, including, "The Lord God is subtle, but malicious he is not." The only alternative in the future that is really "open" to us is the one that actually occurs, but since we don't know in advance which it will be, we are forced to choose among various courses of action. If we don't make a choice, we will probably end up not doing anything! We might not get out of bed in the morning. Or we might end up playing golf all the time. (It's not such a bad life; I know a few people who wouldn't mind trading places with Tiger Woods.) Incidentally, it is not as well known that Einstein later added, "I have second thoughts. Maybe God *is* malicious." Please keep this sentiment in mind as we explore why scientists have allowed themselves to believe that the lack of perfect design in nature is proof of Darwin's theory of evolution.

The fact that we don't know what the future will bring is an important and absolutely essential part of the way things have been designed. If we knew in advance that we were going to break a leg running the New York City Marathon, how many of us would choose to make the effort? Before the race is run, however, the future appears as limitless as our imaginations allow. We may even have hopes of winning! If everything is predictable in principle, things would be very boring (except in places like a casino or the racetrack). Much activity would revolve around refining our measurements and trying to make better predictions. Having irreducible human uncertainty about the future is what makes things interesting and keeps us guessing. We know that we cannot know what will happen, but we realize that only one outcome is possible.

Very few of us go about feeling that we are manipulated by unseen forces; to the contrary, most of us find that there is nothing stopping us from doing what we want. We make choices all the time and exert efforts to fulfill our desires. The fact that the future is not "open" poses few practical problems. Think of the issue in another way. In the view of reality I propose, you have been living in a predetermined world all along. You never have had any "free will," despite what you may have thought. But even though your choices may not have been "free," have you ever felt like a mere puppet? Have you ever felt a lack of "freedom"? Of course not.

Before moving on, let me reiterate Dennett's words about actualism. As we saw, he tells us that he is "prepared to assume that actualism is false" despite the fact that he "can't claim to prove it, if only because the alternative would be to give up and go play golf or something." Quite obviously, Dennett rejects the "dread hypothesis" not because of logic but because of psychology and emotion. He offers no proof that actualism is wrong, but he is very uncomfortable with the implications. He doesn't want to entertain the notion that he might not be in control, so he dismisses actualism by fiat.

5

Cracking Nature's Code of Law and Order

> Everything is determined ... by forces over which we have no control. It is determined for the insect as well as for the star. Human beings, vegetables, or cosmic dust – we all dance to a mysterious tune, intoned in the distance by an invisible piper.
>
> *Albert Einstein, Saturday Evening Post, 1929*

I said earlier that we can know through the use of memory that the past could not have been otherwise, and that this leads to the unequivocal conclusion that the future is completely predetermined. The outcome of events such as the toss of a coin is, in effect, rigged. The view that events can occur in only one way is, of course, in direct opposition to currently accepted scientific thinking. A look at the popular and academic literature shows that the idea of determinism has long gone out of fashion; chance, complexity, chaos, adaptation, emergence, and contingency are all the rage these days. Some scientific theorists hold that the universe is nothing but an accidental and extraordinarily lucky fluctuation; some maintain that chance and randomness rule the world. Steven Weinberg is quoted as saying that "we appear just to have been winners in a cosmic lottery." String theorists and others occasionally posit that we inhabit one of many multiverses (even though the word "universe" has traditionally meant all that is), partly because such a notion (a variant of the anthropic principle) helps explain why the universe is hospitable to life as we know it. Notions of chance, disorder, and entropy have risen to prominence and have found their way into all corners of modern thought.

Sir Arthur Eddington, writing in *The Nature of the Physical World*, sums up the matter as follows: "A common measure can now be applied to all forms of organization. Any loss of organization is equitably measured by the chance against its recovery by an accidental coincidence. The chance is absurd regarded as a contingency, but it is precise as a measure. The practical measure of the random element which can increase in the universe but never decrease is called *entropy*."

In this chapter, we will examine some of the most important principles of modern science, shedding new light on outstanding questions and reevaluating others that would seem to have stood the test of time. Despite the aura of precision and authority that is bestowed upon much scientific thinking, we will see that experimental evidence is in short supply for many important theories. Scientists may have convinced themselves (and many of the rest of us) that their views are based upon solid facts and verifiable data, and they may take enormous pride in their hard-nosed approach to reality, but the foundations of their world are built upon the most questionable of assumptions. In no way can they claim that they approach nature without prejudice. The bias they bring to nature is neither malicious nor unreasonable, but it is still bias.

I am not interested in simply pointing out that the scientific world view contains the highly speculative, anti-factual belief that the "possible" could be the "actual." Using the final knowledge given to us that the future is predetermined, we can be absolutely certain that many scientific assumptions happen to be false. Certainty may not be attainable within science, but it is attainable nonetheless. To make genuine progress, we will have to go beyond science. It turns out that much of what passes for scientific truth is only illusion. All kinds of scientific statements about reality are subject to revision, if not outright elimination.

How many times have you heard something to the effect that nature prefers "disorder over order" or that "randomness continually increases" as time goes by? One physicist tells us that the tendency of children to make a mess of their rooms is an example of this principle in action; others point to the unlikelihood of a shuffled deck of cards returning to its original out-of-the-box, fully ordered pattern when shuffling. Given the presumed applicability of the second law of thermodynamics over a very wide range of phenomena, it may seem impossible that the tenet of nature's preference for "disorder" is based not on experimental evidence but on the prior assumption of "randomness" instead.

Yet this is exactly what we find when we probe the origins of the concepts of entropy and disorder. Eddington once spoke of "the exaltation of the second law" and maintained that the law of entropy increase holds "the supreme position among the laws of Nature." Considering this kind of reverence and the fact that we will be rethinking ideas that have been with us since Ludwig Boltzmann and the nineteenth century, unraveling the mysteries of entropy and dispelling the surrounding confusion will be no small task.

Nevertheless, it is essential that we dissect and deconstruct the origins of the notion that there is a mathematical abstraction called "disorder" that continually increases in nature. (Note that the "disorder" discussed here is a mathematical concept, not the same as qualitative notions of "messy" or "jumbled up." I am perfectly willing to admit that my 4-year old son's room looks disorderly, but this is quite different from saying that nature possesses a characteristic known as disorder.) The second law plays an important role in every corner of science and has a critical bearing on our understanding of natural processes, time, direction, and consciousness, to mention only a few areas. Only by way of detailed examination will we be in a position to understand where science has gone wrong.

In practically all explanations of probability theory, experiments involving the tossing of coins and dice are discussed to illustrate that nature behaves in a "random" manner. The scientist, even before entering the laboratory, possesses a theoretical model that is based on the supposition that there are events in nature that can be called "equally likely." In its most simple form, this assumption states that when a symmetric, or fair, coin is tossed, both heads and tails are "equally likely" to appear. The "equally likely" postulate supposedly gives scientists the authority to make all sorts of pronouncements about nature and reality, especially those pertaining to the notion that nature somehow prefers "disorder over order." (There is even an entire book called *Reality's Mirror: Exploring the Mathematics of*

Symmetry that is more or less based on the notion of "equally likely.")

Yet there is not the slightest evidence, physical or otherwise, to support the view that "disorder" is a property of nature, let alone that the passage of time brings about an increase in "disorder." It is often contended that there is experimental evidence for the scientific viewpoint that "chance" plays a part in the fundamental workings of nature, but we will see that there is none. Only those who are ignorant of the underpinnings of science state that nature has been shown to be ruled by "chance" and "probability." It is unfortunate that extraordinarily large numbers of scientists fall into this category.

When a fair coin is tossed several times, we observe that some of the time it lands heads, some of the time it lands tails. In the long run, we find that it lands heads about 50% of the time, tails about 50% of the time. Does this observation about the long run in any way lead us to the conclusion that on any single toss "we must give equal likelihoods for heads and tails," as Feynman asserts? Not in the least.

If we think carefully about a specific toss that is about to be made, all we can say is that one of two future states (heads or tails) will occur. We have no actual evidence to support the notion that these states are "equally likely" or "equally probable." This view of the event is simply an assumption based on long-term frequencies. The assumption is of such considerable importance that scientists and philosophers have even given it a formal label, "the hypothesis of equal *a priori* probabilities." From my experience, I will give you serious odds that most scientists are not even aware of its existence. In *The Canon: A Whirligig Tour of the*

Beautiful Basics of Science, Natalie Angier writes, "You have, of course, a 50 percent chance of tossing a head (or a tail) with each throw—in other words, a probability of 0.5." This "knowledge," an absolutely essential part of the canon of modern science, is presented as fact (notice the use of the words "of course"); there isn't the slightest hint that such a view is only an assumption. (Generally speaking, I have no problem with assumptions, but they should be identified as such.) To be fair, Angier is a science writer, not a professional scientist.

The relatively few scientists who have even thought about the subject have long argued that the "equally probable" hypothesis is "reasonable," "logical," and "persuasive." The perfect symmetry of the coin, the absence of a measurable differentiating force, and the impartiality of the tossing method all seem to adequately justify the scientist's belief that nature is "indifferent" as to which way the coin will fall. An "indifferent" nature doesn't care about what happens, hence the notion that both future states are "equally likely." We can observe through repeated experiment that the frequency of heads and tails is about the same over time. Therefore, on any given throw, shouldn't both "possibilities" have the exact same chance of appearing? Shouldn't a fair coin and a fair tossing method give us a fair result?

In addition to the rationalization based on the coin's symmetry, the "equally probable" hypothesis is said to be subject to a measure of "*a posteriori* justification." Without going into great detail, the thought here is that because the long-run predictions based upon the hypothesis are accurate, the hypothesis, which pertains to the single event, must be correct. For example, one would predict that in tossing three coins simultaneously, the appearance of all heads would happen about 1/8 of the time. Similarly, four coins would yield all heads about 1/16 of the time. Five coins would yield all heads about 1/32 of the time. These are obviously only the most simple of examples. The confirmation of such predictions through numerous trials helps strengthen our confidence in the validity of the hypothesis. This kind of justification may be roundabout (and it is perfectly acceptable within the rules of science) and seems to bolster the scientist's case.

However, after all our talk about the "logicalness," "reasonableness," and "persuasiveness" of the postulate and the accuracy of predictions that are based on it, the "equally likely" or "equally probable" concept remains only an assumption and not an empirical observation. No experiment has ever verified it (which is why it is called a hypothesis), since it concerns the *individual* trial. Predictions based on it concern average results of a *large number* of trials. In other words, scientists don't predict the individual event and therefore cannot confirm the hypothesis directly. They can't ask a coin if it is "indifferent." They can't ask a coin that came up heads, "could you have come up tails?" Well, they can ask, but they won't get an answer.

Richard C. Tolman is one scientist who is careful to point out that the hypothesis is taken for granted in statistical mechanics; he writes that it "is introduced at the start, *without proof,* as a necessary postulate" (emphasis his). In *The Principles of Statistical Mechanics*, a tome of 661 pages, Tolman spends several pages in various sections discussing the assumption and its origins. Tolman has obviously thought a lot about the philosophical underpinnings of statistical mechanics (which cannot be said of many of today's textbook writers). He is always careful to point out the absence of direct experimental testing regarding probability. (Tolman was no minor scientific figure, by the way; Feynman was the Richard Chase Tolman Professor of theoretical physics at the California Institute of Technology. Tolman was a leading scientist in his day and was cited by Linus Pauling as one of two major influences. There is even now an annual chemistry prize called the Tolman Medal given in his honor.) Other less thoughtful writers (and we will look at the work of a few of them later) claim that there is actual empirical evidence for the notion that there are "equally likely" outcomes for the coin toss, die throw, and similar events, but such allegations are simply the result of unsound and, for those who pride themselves on their rigorousness, extraordinarily sloppy thinking.

I once had a discussion with a leading theoretical physicist, and he spent more than fifteen minutes protesting and obfuscating before finally acknowledging that there is no experimental basis to support the idea that on any given coin toss, heads and tails are equally likely. I find the matter quite obvious. And there is always the work of Tolman that one can consult. However, this world-class theoretician, who had a distinguished Ph.D. (of course), had spent more than fifty years believing that his view of such a simple and elementary event was based on solid experimental evidence. He wasn't about to let go of his beliefs easily, regardless of how wrong they were. No way. The rock-solid, "fact-based" foundation that he had relied upon for firm support throughout his entire career vanished in an instant. Careful thinking showed him that his facts about the coin toss were nothing but assumptions, and he was quite uncomfortable. A (phony) rug had been pulled out from under him, and it was as if he went into free fall.

Let us probe further into the scientific mind-set that underlies the belief that a mathematical abstraction called "disorder" continually increases in nature. For this purpose, we can hardly do better than to consult one of the most widely used physics textbooks of the modern age. In many ways, it is considered the Bible of introductory texts. In *Physics*, David Halliday and Robert Resnick state that "entropy is associated with disorder and the second law statement that in natural processes the entropy of the [system + environment] tends to increase is equivalent to saying that the disorder of [system + environment] tends to increase." To explain what "disorder" means, they consider three examples both qualitatively and quantitatively: the free expansion of a gas; heat conduction; and the stirring of a coffee cup. There is no need to scrutinize all three examples in detail, but it is necessary to look at the essence of the underlying argument. What we will find is that the evidence cited in support of the view that nature tends towards "disorder" doesn't qualify as evidence at all but is rather an interpretation based upon their explicit use of the "equally probable" hypothesis. If the hypothesis is wrong, as I maintain, then their entire point of view is faulty.

In their first example, gas molecules that were originally confined to one-half of a box are allowed to fill the entire box. Halliday and Resnick's qualitative explanation of the expansion goes like this: "By any reasonable definition of the word disorder the system has become more disordered, in the same sense that disorder increases if the litter on one vacant lot is spread over two lots. More precisely, the disorder has increased because we have lost some of our ability to classify molecules." Leaving aside the curious analogy of comparing gas molecules with litter, let us think about the second sentence for a moment. Ignorance is a human property; "disorder" is said to be one that belongs to nature. Yet Halliday and Resnick say that nature's "disorder" has increased because we have lost some knowledge.

I find this a very odd point of view. It is not for nothing that G. S. Rushbrooke once wrote in his *Introduction to Statistical Mechanics* that "randomness and disorder are almost synonymous with uncertainty and ignorance." Why should an increase in our ignorance mean that nature has become more random? But pointing out this strange approach to nature is not by itself enough to challenge the scientific outlook, since science is nothing if not quantitative. And we are by no means engaged in a mere semantic argument. Let us therefore move on to the rigorous realm of mathematics and hard numbers. It is here that we are going to find that the scientists possess a very peculiar *Weltanschauung*.

In their "quite formal" mathematical explanation, Halliday and Resnick state that they will provide a "solid quantitative base" and give "a precise meaning to disorder." They tell us that it can be calculated using the formula

$$S = k \log w$$

where S is the "entropy," or "disorder"; k is a number called "Boltzmann's constant"; and w is the "disorder parameter." (This equation, by the way, is engraved on Boltzmann's tombstone—it is considered that important.) The "disorder parameter" is extremely interesting and is defined as follows:

> "the probability that the system will exist in the state it is in relative to all the possible states it could be in."

Please keep this definition in mind as we move forward. Halliday and Resnick continue, "this equation connects a thermodynamic or macroscopic quantity, the entropy, with a statistical or microscopic quantity, the probability." (It is perhaps worth noting that probability is "conventionally expressed on a scale from 0 to 1; a rare event has a probability close to 0, a very common event has a probability close to 1.")

Let us think rigorously about the above definition of the "disorder parameter," especially the phrase *"all the possible states it could be in."* We know that the system is in a particular state—all we have to do is look. We make a measurement and obtain one and only one result. Where did the scientist come up with the idea that it "could be in" another state? Why did our authors decide that the state the system "is in" has something to do with "probability"?

The answer is of course by assuming, without evidence, that states which did not appear could just as easily have appeared. By assuming, without evidence, that what is not real could be real. By speculating that the theoretically possible could be actual reality. In other words, by assuming CFD, the notion that something that is defined as contrary to fact has a definite reality. As Herbert states, "this CFD assumption, that hypothetical actions would have led to definite outcomes, seems reasonable but it is by its very nature untestable since each event happens only once." (I discuss Herbert's work in more detail in Chapter 7.)

We know the thought process that Halliday and Resnick are engaged in. Since there are large numbers of outcomes over the course of time (the long run), the scientists have decided that an "indifferent" nature allows all of them to be "available" at any given moment. The scientist makes a measurement and gets one and only one reading, but he or she reasons that this result could have been otherwise. Nature gives a very specific result, but the scientist says that she acts randomly. The view put forth by Halliday and Resnick is simply a variation of the "equal *a priori* probabilities" hypothesis and in no way can it be considered impartial. It *presupposes* a specific way of thinking about reality, one which might be summed up as the "ideology of randomness."

If we consider a deck of cards, we have a new state each time we finish shuffling. Imagine that we end up in a state that we will call "Sequence A." To the scientist, all shuffled states are "equally likely," and "Sequence A" is only one of a myriad of states that could have occurred. We could now be in "Sequence B," "Sequence C," "Sequence D," or another of the 52! (factorial) states that arise through repeated shuffling. When you add a second deck to the first deck of cards, the so-called "possibilities" are even larger. Halliday and Resnick even have a homework exercise in which students calculate the change in entropy when a deck of cards is shuffled. However, as I have pointed out, the hypothesis of equal *a priori* probabilities is based on pure assumption. There is no evidence that any one of the many states "available" in the long run could have appeared after the next shuffle.

Returning to the example of gas molecules, Halliday and Resnick tell us that the location of a molecule is "merely a matter of chance" and that there is such a thing as "random molecular motion." (The idea that the molecule's location is due to "chance" would seem to indicate an important scientific point of view, worthy of note and even analysis, yet here our physicists use the term "merely." To them, such a perspective is nothing special or distinctive; chance is simply assumed to be involved in dictating the molecule's location, and no argument or defense is necessary. An inquiring mind might want to know, where did such an idea originate? What evidence is such an assertion based on? We will examine the "chance" mindset in more detail in Chapter 7.) They then use standard theory to calculate the "probability" of a molecule being in a certain location: if a box is divided in two, then the molecule has a 50% chance of being in either half; if a box is divided in four, a molecule has a 25% chance of being in any quadrant, and so on. If we imagine two molecules in a box that is divided into two halves, then the "probability" that they will both be in the same half at the same time is 25%. Halliday and Resnick explicitly use the assumption of "equally probable states" as the basis for concluding that the direction of natural processes is "determined by the laws of probability."

Since the authors start by assuming that the location of a molecule is random, it is easy to see why they think that there is more randomness present after the gas is permitted to expand. After all, once the volume has doubled, the molecule has a larger area to roam around in, and they cannot pin down its location as easily. (Dividing a box into thousands or millions of artificial cells makes the math more complicated, giving rise to all sorts of interesting calculations from the mathematician's standpoint. To many who love quantification, the more numbers and calculations, the better.) Under the scientific interpretation, the molecule would appear to be more "free" after the expansion, resulting in an increase in "disorder."

But it is, or should be, apparent that there is no evidence that the molecule could be in any location other than the one that it is actually in either before or after the change in volume. There never was any proof in the first place that the molecule had a 50-50 chance of being in either half of the box. The "solid quantitative base" that Halliday and Resnick promise us turns out to rest squarely upon a proposition that is completely unsupported by experiment. One might call this quantitative base a metaphysical belief, as there is no physical evidence that can be cited on its behalf.

I am by no means interested in simply questioning the core scientific assumption of a random nature and proposing that it might be incorrect. I propose to go much further. It turns out that the "pure chance" view and the assumption of "equal *a priori* probabilities" (the latter of which is "absolutely fundamental for statistical mechanics," as Walter J. Moore insightfully observes in *Physical Chemistry*) can be known to be completely erroneous. The hypothesis may be entirely "reasonable" and wholly "logical," but it is still unequivocally wrong. As I have already indicated, we can know through the use of memory that the past has occurred in the only possible way and that the future is completely predetermined. The future is not left to chance, accident, or anything of the sort. The coin or die is rightly presumed to be symmetrical, but on any given trial, only one outcome is possible. As I stated before, the only outcome that *can* occur is the outcome that *will* occur. The *possible* and the *actual* are one and the same. Another way of putting it is the following:

> Fair coins yield unfair results. Nature is not indifferent.

We have no way of knowing in advance what will occur, but it is false to think that all of the outcomes that we witness in the long run have the "potential" of occurring at any given moment. Take a deck of cards, shuffle it a few times, and deal a few bridge hands. The card patterns that you see (whether perfectly ordered or one of a multitude of jumbled assortments) are the only ones that could have appeared. There is nothing "random" about the specific result, as it is the only one that was actually possible. The mathematician is very good at calculating "odds" and will tell you the precise "probability" that a player will receive all cards of the same suit, but this is a wholly flawed way of looking at the situation. If you received a perfect bridge hand, it was the only hand you could have received. The odds against it happening were precisely zero, not some meaninglessly large number. (Incidentally, card shuffling is often used to illustrate qualitatively that nature tends towards disorder. If we define 99.999% of sequences as disorderly, it is no wonder that shuffling will appear to produce an increase in disorder. This, however, is not the type of quantitative example used by scientists to demonstrate the existence of disorder.)

Mathematically, Halliday and Resnick's "disorder parameter" w is always equal to the value of 1 (one). The probability that the system will be in the state it is in is always 100%. "Chance" and "all the possible states it could be in" (CFD at work) are mere fictional abstractions that have nothing to do with the real situation. It doesn't matter whether we are dealing with molecules of gas, decks of cards, or a handful of coins. When we conduct an experiment, the measurement we obtain is the only result that we could have obtained. It is false to say that the location of a gas molecule is dependent upon chance. It is false to say that heads and tails both have a 50% probability of appearing or that each face of a die has a 1/6 chance of appearing on the next toss. The scientists and mathematicians have literally figured nature wrong.

Solving the equation put forth by Halliday and Resnick, we multiply Boltzmann's constant k by the logarithm of 1. Since the natural logarithm of 1 is zero, we find that the entropy S has a constant value of zero. More specifically,

$$S = 0$$

meaning that *there never was any "disorder" to begin with*. Disorder (which was a mathematical concept by definition) has no physical reality, as it always has the constant mathematical value of zero. It is merely a figment of the scientific imagination, wholly dependent upon the notion that we could be in a state other than the one we are in.

This "other state" logic [sic] is more reminiscent of science fiction than science, yet the CFD never-never land of "it could be otherwise" lies at the foundation of the second law. Note that "disorder" here is not the same thing as "unavailable energy," although both are called "entropy." My quarrel is not with the proposition that nature has a tendency towards equilibrium, which can be experimentally demonstrated, but with the notion that nature tends towards randomness.

Considering that scientists pride themselves on their "rigorous" and "tough-minded" methodologies, where experimental support rules, I find it highly ironic that a core scientific concept is dependent upon the belief that the "not real" could be "real." Again, I am not suggesting that the formulation of entropy in its original sense based on the work of Clausius needs to be reexamined. My argument concerns the attempt to link entropy with statistics and probability (the line of thinking introduced by Boltzmann).

It is worth noting that the equation $S = 0$ is the mathematical solution to the free will problem, the quantitative equivalent of "you cannot affect the future." The three-symbol equation says that the state the system is in is the only one it could be in (i.e., there is a one-to-one correspondence between the potential and the actual) and that randomness and disorder are constants equal to zero. You are reading my words at this very moment, and you could be doing nothing else. You may think that things could be otherwise, but such is not the case. "What might have been" and could'a, would'a, and should'a completely vanish when one realizes that the actual is the only possibility.

Nothing about my approach is contrary to reason, despite the absurd and ridiculous claims of various scientists that their "equally likely" view is the only logical one or that their long-run predictions, which are borne out by experiment, validate their view of the situation. One of the first principles of science is that different starting points can lead to the same predictions. In reality, the so-called "random sample" is not "random" at all; the one chosen is the only one that could have been chosen, and those not chosen had no chance of being chosen. (Consider the implications for the modern pollster, economist, sociologist, biostatistician, risk manager, or lottery commissioner.) In a casino, we are told that dice "behave randomly," which is supposed to mean that all six numbers on a die have the same likelihood of appearing on the next toss, but this is not the way nature works. *God's dice are always loaded*, and the foundations of probability and statistics are based upon false premises.

I find it quite curious that the scientist approaches nature (prior to conducting a single experiment) with the viewpoint that the motion of molecules is random, that a particle's whereabouts is left to chance, and that an individual object would just as soon be going to the left as to the right. Science starts with the attitude that nature contains disorder, relying upon a metaphysical assumption that has taken on the status of an experimentally verified fact or observation. Since we can know with certainty that only one result is "open" in the toss of a coin, the disorder they perceive has never existed.

I am not going to delve deeply here into why scientists think the way they do, but I sense that Dennett has identified the reason. If the actual is the only possibility, then this puts a constraint on human behavior that most people will oppose. Scientists are people first, scientists second, and about the last thing people want to do is deny their own free will. They want the capacity to shape the future; they want power and control. Scientists would rather believe in a whimsical and capricious chance that gives them apparent freedom of the will instead of an arbitrary, non-accidental, and predetermined design within which there is no real option. As a result, they are going to suggest that nature possesses openness and choice (also known elsewhere as "degrees of freedom"), or the equivalent of free will, so that they themselves can believe in free will.

Scientists tend to balk at the notion that the "miraculous coincidences" among important numbers that allow for life on earth, for example, might be the result of an arbitrary creator (God, to some) rather than random forces. As Feynman says, "nature, as a matter of fact, seems to be so designed that the most important things in the real world appear to be a kind of complicated accidental result of a lot of laws." Feynman and his colleagues see the amazing harmony present in the universe and attribute wildly improbable events (by their calculation) to accidental and capricious forces. They have no problem rejecting the notion of design, which would limit their power.

Scientists, for example, are entirely willing to accept the viewpoint that handedness and other key facts of evolution are due to "accident." In discussing the appearance of right- and left-handed molecules, Feynman, for example, says that "it is easy to believe that the explanation is that long ago, when the life processes first began, some accidental molecule got started and propagated itself by reproducing itself, and so on.... But we are nothing but the offspring of the first few molecules, and it was an accident of the first few molecules that they happened to form one way instead of the other." Notice the phrase "easy to believe" here. Of course scientists find it *easy to believe* in accident as a key explanation; to them the alternative of design is highly distasteful. Scientists want to be in control, and if they can't be in control, it is better that no one be in control. Let me repeat that: scientists want to be in control, and if they can't be in control, it is better that no one be in control. For some, the notion that the universe is the result of a "random fluctuation" is quite comforting when compared to the notion that it was started by design. A designer that is superior to the scientist? Heaven forbid.

6

Quantum Leaps of Faith

> The Heisenberg-Bohr tranquilizing philosophy—or religion?—is so delicately contrived that, for the time being, it provides a gentle pillow for the true believer from which he cannot very easily be aroused. So let him lie there.
>
> *Albert Einstein, letter to Schrödinger, 1928*

I didn't mean to imply in the above discussion that the "chance" view of the coin toss or gas molecule was identical to the quantum mechanical view of nature as "inherently probabilistic." As Feynman notes in speaking about the behavior of molecules, "when probability was first applied to such problems, it was considered to be a *convenience*—a way of dealing with very complex situations" (emphasis his). With molecules or dice, it was assumed that given enough knowledge of initial conditions, one could predict the outcome of an experiment. This meant that classical determinism, which was practically synonymous with prediction, was still tenable, although unworkable in practice. The experiments of quantum mechanics, however, have ruled out the possibility of predicting single events at the subatomic level, because the act of specifying the position of a particle leads to uncertainty as to its momentum and vice versa. In other words, it is impossible, even in principle, to predict the behavior of a single particle.

As Feynman notes when speaking of quantum developments, "the future, in other words, is unpredictable.... That means that physics has in a way, given up, if the original purpose was—and everybody thought it was—to know enough so that given the circumstances we can predict what will happen next." Please reread Feynman's words above. I think that this "*I predict, therefore I am*" mindset goes a long way towards explaining why physicists think that the long-run frequencies they see in quantum experiments translate into "probabilities" for each individual event. A physicist would rather predict *something*, perhaps *anything*, than be forced to admit "the future is unpredictable; I have no idea what is going to take place next." The physicist witnesses one and only one wholly unpredictable result when a single experiment is conducted. But if he or she says that multiple options are open to each electron and that its behavior is governed by a "probability distribution," this helps preserve the original goal of prediction. It is far better to say "the electron has a 50% chance of going through Hole A and a 50% chance of going through Hole B" instead of "I cannot determine which Hole the electron will go through." About the last thing physicists want to acknowledge is that they cannot predict.

But this fundamental limitation on our powers of observation and prediction, wholly unexpected by early twentieth-century physicists, does not support the conclusion that "elementary particles possess a quantum of freedom," as B. K. Ridley asserts, or that "there is probability all the way back: that in the fundamental laws of physics there are odds," as Feynman claims.

Ignorance on the part of physicists does not imply randomness on the part of nature; our irreducible uncertainty is in no way evidence that nature is intrinsically haphazard. Those who support the Copenhagen interpretation of quantum mechanics often take Heisenberg's uncertainty principle to mean that "chance" and "probability" govern the subatomic world, but all this principle says is that we cannot measure to an infinite degree of precision (which is why classical physicists were so shocked, as they had assumed that they would be able to predict ad infinitum). There is a limit as to how far we can go, and we are always bound to be unsure of the single particle's exact behavior.

Over the years, there have been several opponents of the Copenhagen interpretation, notably Einstein, Max Planck, Louis de Broglie, and David Bohm. In general, such physicists attempted to preserve classical determinism and the idea of limitless prediction. However, this desire is unrealistic; there are no "hidden variables" that will allow us to predict the single event as long as Heisenberg's uncertainty principle holds. But even though Newtonian prediction has been abandoned at the subatomic level, Einstein was correct in his view that there is "complete law and order" in the world. It is interesting that towards the end of his life, Einstein wrote to Max Born, "even the great initial success of the quantum theory does not make me believe in the fundamental dice-game, although I am well aware that our younger colleagues interpret this as a consequence of senility." Einstein faced considerable scorn from fellow scientists for his steadfast belief that quantum mechanics was "not yet the real thing," but he was right. He demonstrated remarkable intuition when he concluded, "the theory says a lot, but does not really bring us any closer to the secret of the 'old one.' I, at any rate, am convinced that He is not playing at dice."

In discussing the so-called "validity" of statistical quantum mechanics, let us call upon Tolman again. He asserts that "the methods have to be based, on account of their statistical character, on some hypothesis as to the *a priori* likelihood of different possibilities." The connection between this assumption and the one cited by Tolman in the context of thermodynamics (*"without proof"*) is evident, and he discusses their similarities at length. As in the case of classical physics, it is clear that in quantum mechanics there is no actual evidence for the "equally probable" or "chance" interpretation; it is simply a hypothesis that is accepted at the beginning. However, it turns out that this hypothesis of resident uncertainty in nature is false; using the memory, we can know that the future is already predetermined. Nature knows what will occur, and the future is not "open to all possibilities." Electrons do not possess any freedom. The observation that the same quantum experiment yields many different outcomes over the long run doesn't mean that more than one of these outcomes is possible on the next trial. It does not mean that chance is in operation. The wave function showing all long-run probabilities (frequencies) is said to collapse when a measurement is taken (the potential is transformed into the actual), but the wave function applies only to the long run and has no bearing on the individual event. Yet this important distinction between the long run and the single event has been glossed over by most physicists. Unlike Tolman, the majority of them have been all too willing to advance as "fact" their assumption that nature behaves in a "fundamentally probabilistic" manner.

The quantum approach reminds me of Alford Korzybski's statement that "the map is not the territory…" The quantum map must be statistical in nature as far as long-run predictions are concerned, since no one is going to predict the single event. As a map, the wave function tells us what will happen over time. However, it does not represent the territory on an event-by-event basis. In other words, waves do not apply to the single event. What is taking place in the long run is the measurement of frequencies (how often an event occurs), not probabilities (the odds of its occurrence). Probability is not a property of nature. In the classical world, no Las Vegas casino would bet the house on the single roll of a die. However, casinos have very accurate long-run data that tells them what to expect over time, and they use this knowledge to make money. This has nothing to do with probability, however, and everything to do with frequency.

I was recently struck by a statement made by physicists Nathan Spielberg and Bryon D. Anderson, the context of which was Heisenberg's uncertainty principle. In *Seven Ideas That Shook the Universe*, they write as follows: "according to the positivist mode of thought, if we cannot measure, we cannot know or predict, nor can nature know or predict." (Another often-used phrase is "nature herself does not even know which way the electron is going to go.") Think about that grand pronouncement for a moment or two. The positivists tell us that since we are ignorant, nature herself must be ignorant. Where is the logic here? Is there not a *non sequitur* in this line of thinking? In what way does the irreducible fact of our ignorance about the future lead to the conclusion that nature cannot "know or predict"? Scientific arrogance and egocentrism have been documented by scientists and nonscientists alike, but I find this to be an extreme case! There is no sound reason for projecting our uncertainty onto nature and maintaining that she is also uncertain. (Freud would have a field day here—a psychoanalysis of science and scientists will turn up many interesting observations.) It's like saying, "if we are in the dark, nature must also be in the dark. Uncertainty is a shared property."

This line of thinking brings to mind Eddington's portrayal of scientists as being surprised "that Nature should have hidden her fundamental secret successfully from such powerful intellects as ours." I think there is an unwritten belief shared by many scientists that nature can be no smarter than they are. They like to think of themselves as superior to nature. Anyone who has been around scientists very long knows firsthand how supercilious and pompous they can be as they pursue one of their primary goals, the control and domination of nature. (I am reminded of an engraving etched on an engineering building at the University of Wyoming: "STRIVE ON—THE CONTROL OF NATURE IS WON, NOT GIVEN.")

Positivists generally contend that scientific assertions are meaningless unless they can be directly verified by experiment. Whatever the merits of such a view, it is clear that the positivists haven't been paying close attention to their own assumptions. A strong case can be made that CSICOP members (Committee for the Scientific Investigation of Claims of the Paranormal) have failed to identify sloppy thinking and false logic on the part of many mainstream scientists, including some within their own ranks. Originating at the University of Buffalo and Caltech, this group sees itself as the "science cops," waging battle against "pseudosciences" such as intelligent design, creationism, channeling, pyramid power, talking to the dead, ESP, UFOs, and spoon bending. For the most part, the CSICOPs go after easy targets, some of them on the (lunatic) fringes. One expects that the cops would devote some attention to scientists who falsely represent pure assumptions as facts. For example, Mayr's canonical defense of Darwin—five Facts and three Inferences—conveniently ignores the question of how variation arises. As we have seen, Mayr assumes that variation is produced by chance, but this is utter speculation. Yet his Facts and Inferences are said to show the correctness and airtight logic of Darwinian theory.

I find it extraordinarily ironic that Sagan, the champion's champion of hard evidence, was one of the masterminds behind the search for extraterrestrial intelligence (SETI), an endeavor that is based on assumption and entirely devoid of any evidence whatsoever (so far). If we are here by design instead of accident, much of the probability-based thinking behind the Drake equation and SETI falls apart. As I see it, odds are exceedingly high that ET ain't never going to phone home and that the SETI endeavor, no matter how noble, is a colossal, even astronomical, waste of time, money, and energy. The global "hot lines" established in eager anticipation of ET's earth-shattering call have been deafeningly silent for decades.

Considering that the positivist approach has been a driving force in science for many years, it is no wonder that the vast majority of scientists have erroneously decided that nature's "inherent randomness" has been demonstrated in the laboratory. I cannot resist a minor digression here. Einstein is credited with introducing positivism into science with his theory of relativity.

He later abandoned his early approach. When reminded by a friend that he had initiated the use of positivism in science, he replied, "A good joke should not be repeated too often."

Examples abound of scientists who erroneously think that experiments have shown nature to be inherently probabilistic. Hawking, for example, says that "all the evidence indicates that God is an inveterate gambler and that He throws the dice on every possible occasion." Fritz Rohrlich, discussing Alain Aspect's groundbreaking work in quantum mechanics conducted in the early 1980's, writes that "the probabilistic nature of quantum mechanics has now received new and strong experimental support." He proceeds to talk of "various possible outcomes" for the single experiment, telling us that "which of these values will occur is precisely specified by a probability distribution." Rohrlich is apparently unaware and unconcerned that this view of the situation is entirely dependent on a hypothesis that cannot be supported experimentally. Like so many others, Rohrlich has incorporated "equally likely," "it could be otherwise," and CFD into his thinking to the point that he is not even aware of their critical functions. What the experiments show is frequency, not probability.

More recently, Brian Greene writes in *"One Hundred Years of Uncertainty,"* an op-ed piece in *The New York Times*, that "a quantum calculation lays out the odds that are all roughly comparable that the electron will be in a variety of different locations—a 13 percent chance, say, that the electron will be here, a 19 percent chance that it will be there, an 11 percent chance that it will be in a third place, and so on. Crucially, these predictions can be tested." Unfortunately, Greene's notion of testability (the bedrock of science, as we have seen) is, in this context, entirely false and exceptionally misleading. The prediction that the electron has one chance of being here and another chance of being there concerns the individual trial and *cannot* be tested. As we have seen, when you make a measurement, you obtain one and only result. There is no evidence to support the notion that odds are involved.

Physicists like to make much of their experiments, but they do not demonstrate in the laboratory that during the next trial the electron is free to choose among the various outcomes that are seen in the long run. They do not demonstrate that nature could have done otherwise in a given situation. Insofar as Greene's predictions are borne out by experiment, they concern the long run, not the individual trial. This is quite clear when Greene tells us, "Take an enormous sample of identically prepared atoms, measure the electron's position in each, and tally up the number of times you will find the electron at one location or another.... if quantum mechanics is right, in 13 percent of our measurements we should find the electron here, in 19 percent we should find it there, in 11 percent we should find it in that third place. And, to fantastic precision, we do." Measurements are based on multiple trials and give us an observed frequency distribution. Odds, however, are about what will take place next (i.e., the odds that an electron will be here, the odds that it will be there), and, as I noted above, there is no evidence whatsoever that odds have anything to do with the situation.

Greene writes, "Can it really be that the solid world of experience and perception, in which a single, definite reality appears to unfold with dependable certainty, rests on the shifting sands of quantum probabilities? Well, yes. Probably. The evidence is compelling and tangible." It is clear that Greene has never read (or perhaps I should say, understood) Tolman, who said, "the methods have to be based, on account of their statistical character, on some hypothesis as to the *a priori* likelihood of different possibilities.... an entirely similar point of view as to the validity of statistical mechanics is adopted, for the purposes of this book, in the quantum as in the classical case." Tolman tells us quite clearly that the postulate regarding probabilities for the single event is assumed *"without proof"* (his emphasis, as I noted earlier). The postulate (conjecture, hypothesis, premise) is considered to be self-evident and is taken for granted; it is an *untested* assertion or assumption. (You can look up the definition for yourself.) In no way is there actual proof of an inherently probabilistic or random nature.

I am not quite sure what "compelling and tangible" evidence Greene sees to justify his view that the world of everyday experience "rests on the shifting sands of quantum probabilities." However, I think it is apparent that his understanding of the situation stems from the false and erroneous melding of the long run and the individual event. Greene thinks that accurate (long-run) quantum predictions validate the physicist's view of the individual trial, but this view is based on assumption, hypothesis, and belief; it does not derive from experiment, test, or measurement. Greene's so-called "evidence" would not pass muster at a conference of logicians and would be laughed out of court by a competent trial attorney. "Objection! Inadmissible by virtue of its nonexistence," Perry Mason might bellow during cross-examination. And Hamilton Burger would look bewildered as usual.

Let us examine Greene's use of the word "crucially," which is the adverb of "crucial," in his sentence "Crucially, these predictions can be tested." Two definitions of "crucial" are the following: (1) of extreme importance; vital to the resolution of a crisis; "a crucial moment in his career"; "a crucial election"; "a crucial issue for women" [syn: important] (2) of the greatest importance; "the all-important subject of disarmament"; "crucial information"; "in chess cool nerves are of the essence" [syn: all-important(a), all important(p), essential, of the essence(p)]. It is, of course, "of extreme importance" or "of the greatest importance" for physicists to think that their predictions can be tested; otherwise they themselves would admit that they were engaging in something akin to metaphysics or, as they condescendingly put it, mere philosophy. Remember the MIT physicist's withering dismissal of Huston Smith I cited earlier: "Think of you? We don't even bother to ignore you." The last thing most physicists would want to be pursuing is speculation and untestable theory. If only to preserve their own sense of self-esteem and lofty position in the pecking order, they want to think that they are engaged in solid, experiment-based, fact-finding. (Let's not talk about string theory, however; the recent critiques by Lee Smolin and Peter Woit delve into the untestable and even unscientific assertions of string theorists quite eloquently.)

Yet it is quite clear that the physicist's claim that the electron could go in any number of different directions during the next experiment is not subject to testing. Sagan tells us that "propositions that are untestable, unfalsifiable are not worth much." I wouldn't necessarily go that far, but this notion could easily be applied to the quantum predictions cited by Greene (who is merely repeating the party line that has been in place since the early days of quantum theory). Although Greene, to his credit, also notes that the predictions are of a "statistical" nature, he promotes the (false) view that the "universe evolves probabilistically." Greene is one of the world's most visible physicists, and I suggest that the vast majority of his *Times* readers are going to accept his contentions about nature's randomness at face value. The general layperson doesn't stand a chance of realizing that accurate long-run statistical predictions do not demonstrate that individual quantum events are inherently probabilistic.

One can argue that Greene and his colleagues can be forgiven for their confusion and imprecision—probability indoctrination begins at a very early age, and the underlying hypotheses are not mentioned. It is easy to go from the qualitative elementary school concept of "what might happen" when a coin is tossed to the quantitative college level assertion that "there are equal probabilities for heads and tails" without pausing too much to think. I do not fault Greene for employing a hypothesis; I fault him for taking a hypothesis and presenting it as a fact that has been confirmed through testing, or what he calls "decades of painstaking experimentation." All the experiments in the world don't show that nature is intrinsically random. When it comes to probability, I have encountered numerous scientists who are unable to see the distinction between hypothesis and observation. They are so wrapped up in their measurements that logic fails them.

At the risk of repetition, the confirmation of quantum mechanical predictions does not in any way support the view that nature herself is probabilistic. It demonstrates that we can make accurate long-run forecasts, but it gives us no insight into the single event. The notion that the physicists are testing their predictions is pure imagination, and the belief in odds and inherent randomness, where "probability" dictates what will occur next, is nothing more than an assumption. As Herbert and others have noted, the typical physicist is of the opinion that as long as the math works, one need not be too concerned with anything else. I recently read that MIT's academic atmosphere is "brutally competitive," but much of what took place in the way of learning when I was an undergraduate revolved around memorization and simplistic problem solving, not particularly deep thinking. What Nassim Nicholas Taleb calls "erudition" in *The Black Swan*—genuine intellectual curiosity—was not a priority at all. Although no one is expecting comprehensive liberal arts training at an institution such as MIT, focusing too narrowly without an examination of first principles can lead to gross misperceptions about what nature actually tells us. Solving puzzles and inventing explanations for experimental results does not necessarily translate into knowledge or understanding.

Furthermore, it is intuitively obvious that the "one possibility" view of the single coin toss or quantum event that I champion yields predictions that are the same as those obtained from the "probabilistic" view. Remember, the only kind of predictions made are about the long run. (As I said before, no scientist is going to predict what will happen next, just as no casino will bet the house on a single roll of the die.) This means that the agreement between observation and the mathematical theory of chance isn't a problem for those who, like myself, reject randomness. It is commonly known in science that an incorrect hypothesis can yield accurate and useful predictions, and this is what has happened in the case of quantum mechanics. But the fact that the long-run scientific predictions are correct doesn't mean that there is genuine understanding about the single event. What is at issue is the underlying view or philosophical outlook that leads to the predictions, not the predictions themselves. Neither my view nor the scientific assumption can be proved or disproved by experiment.

Some of you who have wholly accepted the idea of chance may say that the "burden of proof" lies on my side as challenger, but consider the following: I contend that the result that *did* occur was the only result that *could have* occurred; the adherents of chance say that other results *could have* occurred. I maintain that the real, the actual, the observed measurement was the only possibility. Scientists bring unreal, imaginary, and theoretical states into play and say that they were possible. They say that chance allows all possibilities. They tell us that even though event "A" really took place, event "B" could just as easily have occurred. They contend that a coin that came up heads could have come up tails. They say that a die that shows the number "6" could just as easily show a "1," "2," "3," "4," or "5." They allege that an electron that went to the left could have gone to the right. Accident, randomness, chance, luck, and the hypothetical abound in science.

Whose is the weaker position? Quite frankly, I don't think that there is any doubt. CFD, after all, is an acronym for

Contra-Factual Definiteness, which means being definite about something that is *against the facts*. CFD stands for *anti-factual, anti-reality*. If we go by facts and observations (remember Sagan), the idea that "it could be otherwise" is very, very weak, since it is mere speculation as well as excess baggage. I predict that the vast majority of you will agree that the scientific never-never land view is by far the more tenuous, as it indicates the desire to deny what actually happened. We are always in the state that *did* occur and are never in the state that *did not* occur. We have never experienced a negation of the real. As far as I can tell, the application of Occam's razor would favor my no-CFD approach, as it is simpler and doesn't bring the unreal into play. Once you open a can of (unreal) worms, there is no telling where imagination and fantasy will take you.

The old saying "a bird in the hand is worth two in the bush" will give you an idea of the relative weight placed upon the real versus the potential. The math makes no sense (how could one bird be worth the same as two birds?) unless one realizes that the comparison has to do with two completely different states of existence. One is real and the other is imagined. The real takes precedence, as you might expect. Have you ever passed up an empty parking space, hoping to find another that is closer to your destination, only to be denied the hypothetical space as well as the real one you rejected? If not, your parking abilities are extraordinary.

When have we ever had grounds for belief in "alternative realities," "ghost worlds," "frozen accidents," or "parallel universes"? (I don't even want to get into branes and multiverses, which I find bordering on science fiction. Paul Davies has written,

"The multiverse theory is increasingly popular, but it doesn't so much explain the laws of physics as dodge the whole issue.") I recently read a critique of much of modern physics, where the author said the question is not, "Where can I let my imagination take me?" but rather "What is Mother Nature trying to tell us?" It is easy to imagine trillions upon trillions of alternative worlds (the numbers have gotten bigger since Sagan's time, and the mind boggles with the possibilities), each a variation on our own, but I see no actual evidence to support such speculation. These concepts are very fashionable in physics today and keep many engaged in abstract mathematical calculations and flights of fantasy, but there isn't one iota of laboratory evidence to support them. I realize that I may not be able to bring forth experimental verification any more than the Copenhagen adherents can, but then again, I don't rely on *what did not occur*, on *what was not seen*, as a fundamental building block. I don't rely on CFD. Unlike the scientists, I don't build the unreal into my system.

The answer to the Schrödinger's cat problem, which eluded Einstein and his colleagues, is that it is a false problem. It is not really a problem at all, because the experimental apparatus proposed doesn't exist in nature. There is no such thing as a radioactive particle that has a 50-50 chance of decaying in a given time frame. We do not know what the radioactive particle is going to do, but its future is completely predetermined, not left to chance. Even though recent experiments indicate that some atoms can spin both clockwise and counterclockwise at the same time, this does not mean that both directions are possible when a measurement is made. The uncertainty regarding which direction will appear has to do with our ignorance and is not part of the experimental setup. There is uncertainty on the part of humans, but not on the part of nature.

In summing up the traditional view of quantum phenomena, Eyvind H. Wichmann states in *Quantum Physics* that "the important feature of the thermal motion in a system is that it is, from our standpoint, a random motion. It introduces an apparent element of chance in the behavior of the system, as observed by us. We can say that the thermal random motion is 'noise in the symphony of pure quantum mechanics.' And we can add that often the noise is so loud that the music cannot be heard." I would like to propose that Wichmann has got the music and the noise backwards. There is nothing "random" or "chancy" about the thermal motion; at any time, a particle is moving in the only way it could possibly be moving. Wichmann's apparent element of chance, wholly assumed from the start, has no physical reality.

This is not the time to explore in detail why Wichmann and his colleagues describe nature as being "random," "chaotic," "capricious," "disorderly," "fickle," "whimsical," "aimless," and "indifferent" (coincidentally, such words tend to be full of negative connotations), but briefly, I suspect that it has to do with their inability to predict. It is not hard to see that once quantum mechanics demolished the cherished, but flawed, classical notion of unlimited predictability, it became acceptable to go from a convenient chance to an inherent one. The logic (if you can call it that) goes something like this: "If nature is predictable, then it is governed by law and order. But if nature is not predictable, then it must be governed by chance. Uncertainty on the part of humans means uncertainty on the part of nature." Hawking, for example, begins a lecture called "Does God Play Dice" with the following,

"This lecture is about whether we can predict the future, or whether it is arbitrary and random." Prediction and randomness are the only two alternatives offered. Hawking thinks that our inability to predict means that nature is random. However, there is no reason in the universe why a non-random, deterministic nature should be predictable to physicists. Determinism and predictability never should have been equated in the first place.

7

Sagan's Baloney Detection Kit Applied

Nowadays whenever enthusiasts meet together to discuss theoretical physics the talk sooner or later turns in a certain direction. You leave them conversing on their special problems or the latest discoveries; but return after an hour and it is any odds that they will have reached an all-engrossing topic—the desperate state of their ignorance. This is not a pose. It is not even scientific modesty, because the attitude is often one of naive surprise that Nature should have hidden her fundamental secret successfully from such powerful intellects as ours.

Sir Arthur Eddington,
The Nature of the Physical World

In *The Demon-Haunted World: Science as a Candle in the Dark*, Carl Sagan introduces a "Baloney Detection Kit," a toolkit for ferreting out careless thinking, false logic, and highly suspect claims. I find his analysis quite useful. Sagan takes great pains to stress the importance of experimental evidence in science with repeated statements such as the following: "propositions that are untestable, unfalsifiable are not worth much" and "claims that cannot be tested, assertions immune to disproof are veridically worthless, whatever value they may have in inspiring us or in exciting our sense of wonder." Let's take a short quiz. Put your thinking cap on and apply Sagan's toolkit to the following examples. As you read through the seven selections, ask yourself the following questions:

1. How sound is the logic?
2. What evidence supports the conclusions?
3. Is the statement subject to testing and/or falsification?
4. Are there any hidden assumptions at work?

At the end of the chapter, we will assess whether or not the selections pass Sagan's tests for sound science. If not, we should be prepared to call them baloney.

Example 1
"The human race has an incredible amount of experience to assure the statistician that an honest well-made die, properly thrown, will yield its six faces with (almost) exactly equal frequencies. That is, any one of the six numbers is just as likely to come up as any other. Therefore, this method, limited only by the unavoidable small imperfections of the die, *guarantees* to each of the six persons the same opportunity to be chosen" (their emphasis).
J. L. Hodges, Jr., David Krech, and Richard S. Crutchfield, *StatLab: An Empirical Introduction to Statistics*

Example 2
"The tossing of a coin has two possible outcomes and extensive experience has taught us that they are equally probable, so each outcome is assigned a probability of ½."

        Martin Goldstein and Inge F. Goldstein, *How We Know: An* Exploration of the Scientific Process

Example 3
"In the ideal gas the position of a molecule at any time is a result of pure chance."
        Gilbert W. Castellan, *Physical Chemistry*

Example 4
"The fundamental difference between the first and second steps of natural selection should now be clear. At the first step, that of the production of genetic variation, everything is a matter of chance. However, chance plays a much smaller role in the second step, that of differential survival and reproduction, where the 'survival of the fittest' is to a large extent determined by genetically based characteristics."
        Ernest Mayr, *What Evolution Is*

Example 5
"Now, most single accidents make very little difference to the future, but others may have widespread ramifications, many diverse consequences all traceable to one chance event that could have turned out differently. Those we call frozen accidents.... Let's suppose that this conclusion is correct and that the right-handedness of the biological molecules is purely an accident. Then the ancestral organism from which all life on this planet is descended happened to have right-

handed molecules, and life could perfectly well have come out the other way, with left-handed molecules playing the important roles."

Murray Gell-Mann, *"Plectics"*

Example 6
"We are going to die, and that makes us the lucky ones. Most people are never going to die because they are never going to be born. The potential people who could have been here in my place but who will in fact never see the light of day outnumber the sand grains of Arabia. Certainly those unborn ghosts include greater poets than Keats, scientists greater than Newton. We know this because the set of possible people allowed by our DNA so massively exceeds the set of actual people. In the teeth of these stupefying odds it is you and I, in our ordinariness, that are here."

Richard Dawkins, *Unweaving the Rainbow*

Example 7
"Compute roughly how many acts of sexual intercourse occur each day in the world. Does the number vary much from day to day? Estimate the number of potential human beings, given all the human ova and sperm there have ever been, and you find that the ones who make it to actuality are ipso facto incredibly, improbably fortunate."

John Allen Paulos, Innumeracy

Let's consider Example 1. You will have noticed, I hope, the absolutely disastrous flaw in logic shared by our statistics professors. There is a *world of difference* between saying that a die will yield six "(almost) exactly equal frequencies" and "any one of the six number is just as likely to come up as any other." I will call this erroneous equating of the long run with the individual event the Conflation Fallacy. One might say that the statisticians' statements exist in parallel universes, never to be related. The former concerns the long run and can be verified by counting. The latter, a *non sequitur*, is purely an assumption dealing with the single event and has absolutely nothing to do with an "incredible amount of experience." The seemingly authoritative and trustworthy guarantee that the statisticians offer us (notice how they use italics to emphasize their point) is utterly worthless, about as valuable as the guarantee offered by the former makers of the Yugo (a car which is said to have doubled in value after being filled with gas) or the reassuring pronouncements made by Enron executives right before its collapse. Yet I would be willing to bet that not one out of a thousand students reading the passage is going to sense any problem with the logic [sic] employed. After all, our authoritative professors are making authoritative statements and offering us an authoritative guarantee. Why should anyone doubt them? (Observe the commanding word "empirical" in the subtitle of their textbook.) Nevertheless, the statisticians' unconditional guarantee turns out to be a scientific and mathematical fiction, nothing more. It is completely bogus. (Here is a Zen-like koan for your consideration: if the guarantee underlying all of statistics and probability is false, should there be a recall? What about a refund?)

Say that you have just rolled a die and that the result was "6." What "incredible amount of experience" do you have that another number was "just as likely to come up"? What "experience" is there that the outcome could have been otherwise? What evidence is there that other numbers were genuine possibilities? None whatsoever. Let me restate that at the risk of being obvious: there is no evidence whatsoever that numbers that did not appear were just as likely to appear. Yet the belief that an event that did not occur could have occurred is absolutely critical to probability theory and the formulation of the second law of thermodynamics, to say nothing of risk analysis, exit polling, economic forecasting, and lotteries. (Probability-based sports analyses that purport to tell us the probability that a team will win a game according to the outcome of each play as the game unfolds are based on false assumptions and are for the most part irrelevant.)

In Example 2, the Goldsteins share a similar confusion as to what qualifies as evidence. Unlike the clear-thinking Tolman, they engage in the Conflation Fallacy along with our statisticians. This leads them to the false belief that experiment has confirmed the core hypothesis underlying all of probability theory. Long-run statistical data in no way validate an assumption about the single event, as we have seen. Given the title of their book, *How We Know*, one might have expected more precise thinking. If professors teach like this, it is no wonder that their students are clueless about the existence of the hypothesis of equal *a priori* probabilities.

Let's move on to the next example. Using Sagan's requirement of evidence, one would have to question whether Castellan's statement (Example 3) qualifies as science at all. The notion that "the position of a molecule at any time is a result of pure chance" is simply an assumption that results from a world view that is probably unconscious and hidden (remember the chance-colored eyeglasses I mentioned earlier). Evidence? What evidence? Objectivity? What objectivity? Here, the emperor has no clothes and is also unaware of it. Castellan offers not the slightest proof for his crucial assertion, since he has none. He simply declares that his view of the experimental situation is true. One could argue that the scientist's outlook is largely unchanged from that of Democritus and the ancient Greeks, who also believed in the vagaries of chance.

Excuse me for asking, but how could there be evidence to support the notion that "pure chance" is responsible for the position of a molecule? One might view Castellan's assertion as a "weird belief," using the criteria Shermer sets forth in *Why People Believe Weird Things: Pseudoscience, Superstition, and other Confusions of Our Time*. Is it not weird or strange to think that the position of a molecule is due to chance? Where did such an idea come from? The molecule is doing whatever it is doing, and the idea that chance has something to do with the state of affairs is mere conjecture and nothing but conjecture (an uncharitable wag might even call it superstition). Why not say that the molecule is supposed to be where it is? One could argue that such a view is as justifiable as Castellan's. (Or maybe Bigfoot, the Loch Ness Monster, or Casper the Friendly Ghost is responsible.)

Nevertheless, the "pure chance" view of molecular activity is accepted as solid fact by generations of scientists. I remember using Castellan's classic textbook in one of my courses at MIT. When it came time to review notions of probability and chance, the scientific perspective was unquestioned. Students nodded their heads in mute agreement when told that "pure chance" was responsible for the position of a molecule at any given moment in time. (As I mentioned earlier, undergraduates in MIT's science classes are told what to think; there was no questioning of accepted doctrine.) Similarly, Eddington tells us that "each molecule is wandering round the vessel with no preference for one part rather than the other" as a building block for his assertion that "entropy continually increases." The notion that the molecule could be somewhere besides the location it is actually in is the foundation that underlies the contention that nature tends towards disorder.

For some reason, scientists prefer to think that nature allows molecules to wander about governed by chance even though there is no way to confirm such a theory through experiment. They like to think that nature is indifferent as to the molecule's location. If such as view is not falsifiable, shouldn't it be considered a metaphysical proposition or "mere philosophy," as some might say? Scientists long ago decided that there is no difference between the statements "we are uncertain of the molecule's location" and "the molecule's location is left to chance." In short, *human* ignorance leads to the conclusion that *nature* acts in a haphazard, indifferent, and random manner.

As an aside, the assumption that nature is indifferent as to left and right was for centuries taken for granted by scientists. When it was discovered in the late 1950's that nature *does* distinguish between left and right, resulting in the overthrow of the longstanding law of parity, scientists all over the world were dumbfounded and astonished. Many prominent scientists such as Feynman had wrongly predicted that parity would be conserved. (Feynman lost a bet to another physicist.) They had assumed that nature would be indifferent, and they were quite shocked when this turned out not to be the case. Some scientists even went so far as to say that we now needed to give nature more credit. They had thought that dumb nature didn't have the ability to distinguish between left and right, even though they themselves have no problem doing so.

Overnight, an entire way of thinking that was deeply ingrained into the scientific psyche was found to be false and incorrect. As Isidor I. Rabi said at the Columbia University press conference during which the unexpected findings were announced, "A rather complete theoretical structure has been shattered at the base and we are not sure how the pieces will be put together." (An understanding of the downfall of parity is absolutely essential if one is to understand how science works. Martin Gardner's *The New Ambidextrous Universe: Symmetry and Asymmetry from Mirror Reflections to Superstrings* is an excellent starting point.)

To continue our minor digression, a similar belief in the indifference of nature is largely responsible for the thalidomide tragedy of a half-century ago, when about 10,000 deformed babies were born, mostly in Europe. As John Lienhard notes, "thalidomide, which created so many birth defects in the 1950s, is chiral [handed]. One form is a harmless morning-sickness suppressant. The mirror image is an active mutagen. The drug was put on the market with both forms present, and it did untold damage." The body could utilize only one form of the drug, but scientists did not realize this preference until thousands of limbless infants had been born. The scientist's false assumption that nature doesn't care about left and right was a key factor behind a major medical catastrophe whose legacy is still being felt today.

In this case (and in many others), nature was far from indifferent. (Thalidomide has since been found to help fight certain diseases, and its use has been reinstated under strict safety requirements. *Dark Remedy: The Impact of Thalidomide and Its Revival as a Vital Medicine,* by Rock Brynner and Trent Stephens, is a fascinating account of the thalidomide story from its beginnings to the present day.)

Mayr's statement (Example 4) that "everything is a matter of chance" regarding genetic variation is quite similar to Castellan's assertion. Both are bold pronouncements unaccompanied by any evidence whatsoever. Mayr's passage brings to mind the famous statement "Who ordered that?" made by Rabi when the muon was unexpectedly discovered. Who came up with the curious idea that "chance" is somehow responsible for the mutation and variation that we see taking place around us? Show us the proof, the CSICOP crowd might demand if they were to cast a cold, dispassionate, and unbiased eye on evolutionary theory and give it the same skeptical treatment they give to pyramid power, alien visitation, spoon bending, channeling, and the Bible Code.

Mayr's declaration demonstrates an ideological mindset predisposed to the belief in chance. It does not pass Sagan's requirements of experimental results, data, observations, and measurements. It does not qualify as a scientific statement, despite that fact that nearly all scientists consider Darwinian theory to be correct. (By the way, I have no quarrel with the general notion of "survival of the fittest," which, as has often been pointed out, is rather obvious. Thomas Huxley, Darwin's bulldog, thought that the concept is so obvious that he wondered why he himself did not come up with it.)

Let us move on. Gell-Mann (Example 5) believes in "accidents" and tells us that events that did occur "could have turned out differently." There is, of course, no physical evidence to support this viewpoint, yet it forms an important pillar of Darwinism, complexity theory, and much scientific thinking in general. The notion that things "could perfectly well have come out the other way" is, of course, taken for granted by most of us, but it certainly is not subject to experimental verification (and hence not falsifiable). Perhaps it too should not be considered science if we are to use Sagan's rules.

In *Quantum Reality*, Herbert goes into great detail concerning the line of reasoning employed by Gell-Mann and others, which he calls Contra-Factual Definiteness (CFD). Herbert points out that this "untestable" assumption is omnipresent and is hidden in theories such as John Wheeler's "delayed choice" interpretation of quantum mechanics. Herbert discusses choosing among pizzas and placing an order: "In my pizza pie analogy, the CFD assumption means that I take for granted the notion that ordering any kind of pizza other than the one I did in fact order would have resulted in its delivery. This CFD assumption, that hypothetical actions would have led to definite outcomes, seems reasonable but it is by its very nature untestable since each event happens only once. You can order only one pizza this Saturday night...."

In short, CFD assumes that the "possible" could be the "actual." You ordered a cheese pizza but you think you could have ordered pepperoni (or anchovy, or onion, or mushroom, or ham, or salami) instead. You think that a pizza that did not arrive at your door could have arrived, which would now place you in a different reality. (Maybe you could have avoided a case of heartburn.) CFD is, of course, almost a given in our perceptions of reality. We all think that we could have done otherwise and that things could now be other than what they are. I once asked E. J. Squires, author of *Conscious Mind in the Physical World*, why people believed in CFD. His response was, "well, you have to start somewhere." Think about that statement for a moment. You *start* by believing that the real could be unreal. You *start* by believing that you could have acted differently. This is, of course, a very comforting thought, one tied to notions of power and control. But the belief that the unreal could be real has nothing to do with what Sagan would call sound science.

I am not going to pretend to "disprove" the common belief that things could be otherwise, as such a task would be impossible. After all, we are dealing with an assumption that is not subject to testing and proof. (The fact that the assumption is wrong is another matter entirely.) What I want to do here is to point out that scientific assertions about chance and accident are not the result of a "value-free" and "objective" approach to nature but depend instead on the prior assumption of chance, disorder, and randomness.

Let us now consider Dawkins' passage (Example 6), which is an absolutely perfect blend of CFD, the belief in luck, and unwarranted mathematical certainty. Dawkins' world is full of pretend and make-believe and is quite typical of the kind of thinking that characterizes evolutionary biology, where "what might have been" plays an absolutely central role. In Dawkins' world, theoretical possibilities abound ("potential people," "unborn ghosts," and "scientists greater than Newton"), whereas the actual is limited to what we see. The former is said to be vastly larger than the latter. Dawkins says that "we know this because the set of possible people allowed by our DNA so massively exceeds the set of actual people."

Such thinking is pure CFD and is not in any way supported by the facts. Dawkins can calculate as many DNA combinations as he wants (and this theme runs almost ad infinitum throughout his work), but this does not mean that the "possible" is greater than the "actual." Notice Dawkins' highly confident and self-assured use of the phrase "we know." Unless Dawkins possesses supernatural powers unavailable to the rest of us, there is absolutely no way that he could know that the potential could be actual. There is no way that Dawkins could have knowledge that the unreal could be real. Dawkins is simply making the very common and illogical mistake of reifying the hypothetical. One might have expected clearer thinking from such a celebrated and influential figure.

Now compare Dawkins' statement to what T.S. Eliot (who, by the way, studied philosophy with Bertrand Russell) says in *Burnt Norton*:

> What might have been is an abstraction
> Remaining a perpetual possibility
> Only in a world of speculation.
> What might have been and what has been
> Point to one end, which is always present.

Eliot realizes that "what might have been" (Dawkins' "people who could have been here") exists only as speculation and cannot be granted any standing whatsoever in the physical world that we actually experience. In no way can this imaginary world of abstraction be considered knowledge. (Dennett also realizes this critical and obvious distinction, of course, being a trained philosopher.) Dawkins' "we know" is a wholly false and unscientific statement (and skeptics such as Shermer and his fellow science cops would have pointed this out long ago if only they used a little more critical analysis. It shouldn't matter that Dawkins is on the Editorial Board of ***Skeptic***).

We have exactly and precisely no knowledge (zero, nil, zilch) that the potential could be actual. A young musician who is said to have potential might later become a world-class performer in actuality, but potential is always "out there" and unrealized. In fact, Dawkins' views might be characterized as anti-knowledge or anti-fact, since they are without question and by definition *against the facts*. Possibility is anti-actuality. Instead of "we know," phrases such as "we believe," "we posit," "we assume," "we speculate," "we surmise," "we hypothesize," "we theorize," "we suppose," and "we imagine" are closer to the thought process that is taking place in Dawkins' world. But because Dawkins tells us that we can know the unknowable, speculation ungrounded and unchecked by the facts rules the day. Nearly anything is possible in this world of pseudoscience and pseudo-math, no matter how ridiculous (or sublime).

Considering the never-never land quality of Dawkins' reasoning, I am not sure how it passes for genuine science. It doesn't in any way fulfill Sagan's test requiring "experimental results, data, observations, measurements, 'facts.'" It doesn't even pass Dawkins' own criterion for why one should believe in one explanation instead of another. In a letter called "Good and Bad Reasons for Believing," written to his then-10 year old daughter, Dawkins continually stresses the need for "evidence." Yet here there is no evidence, and there can't be when talking about the hypothetical. The facts in this case are the set of "actual people"; the rest ("stupefying odds") is pure conjecture.

The indisputable reality is that you were born in this day and age and are now alive. But according to Dawkins, this shouldn't be the case at all. He tells us that "the odds of your century being the one in the spotlight are the same as the odds that a penny tossed down at random, will land on a particular ant crawling somewhere along the road from New York to San Francisco. In other words, it is overwhelmingly probable that you are dead." However, such a point of view (lottery thinking, where chance abounds) is nothing but speculation (although Dawkins does not realize it).

Recall Dawkins' words about your parents and grandparents that I quoted in Chapter 4, where he claims that "the thread of historical events by which our existence hangs is wincingly tenuous." Presumably, we are supposed to feel quite special at having overcome great odds throughout a long chain of historical circumstances. As Dawkins sees it, any slight and easily imaginable deviation could have resulted in our not being here. Dawkins can certainly believe in his notions of vast numbers of unborn people, highly unlikely events, and the overwhelming probability that you should be dead, but what we have here is interpretation and imagination (wonder), not rigorous measurement, experiment, or evidence. As Dawkins said in another context that might also apply here, "what we need is less gasping and more thinking."

Dawkins has recently stated that he believes that "all life, all intelligence, all creativity and all 'design' anywhere in the universe, is the direct or indirect product of Darwinian natural selection. It follows that design comes late in the universe, after a period of Darwinian evolution. Design cannot precede evolution and therefore cannot underlie the universe." Dawkins has it exactly 180 degrees backward. We are here by design, and the actual animals and the theoretical animals that he maintains "*could* exist" are one and the same. Dawkins posits the hypothetical as being real possibilities, but there is no physical basis for doing so if we are to rely on the actual facts. According to Sagan, "you must be able to check assertions out. Inveterate skeptics must be given the chance to follow your reasoning, to duplicate your experiments and see if they get the same result. The reliance on carefully designed and controlled experiments is key." In Dawkins' hypothetical world of people who don't exist, there is no experiment or evidence. If any science is taking place, it is science fiction.

Dawkins' *The Blind Watchmaker* is subtitled *Why the Evidence of Evolution Reveals a Universe Without Design*, but the evidence does not tell us that events occur randomly or by chance. Dawkins has falsely equated his subjective interpretation of the evidence with indisputable, undeniable proof. As Shermer (following Francis Bacon) rightly tells us in *Science Friction: Where the Known Meets the Unknown*, all evidence must be interpreted. It appears that Dawkins has forgotten to apply this critical point to his own reasoning. The mutation does not say, "I occurred randomly and could have behaved otherwise." The view that variation occurs by chance is, as we have seen, mere assumption.

It is worth noting that the passage from Dawkins I quoted above comes at the very beginning of the first chapter of

Unweaving the Rainbow. With CFD presented as solid knowledge, Dawkins embarks on a long, winding odyssey based on what might have been and what might not have been. His prose may be quite enjoyable (a good read, some might say), but we are in a position to completely reject chance, randomness, and luck (one of Dawkins' favorite concepts) as explanations for our being here. The main line of thinking behind *Unweaving the Rainbow, The Blind Watchmaker,* and another of Dawkins' books, *Climbing Mount Improbable,* is pure illusion. One might even use the word superstition. In reality, the odds against our being here were precisely zero, no more, no less.

One wishes that Dawkins, who is very publicly engaged in the public understanding of science, could produce evidence showing that the potential could be actual, but it can't be done. We are alive in the here and now; the notion that things could be otherwise is not based on evidence. Dawkins makes the imposing claim that Darwinian natural selection is an "established fact," but this ridiculous pronouncement indicates that a leading scientist can be quite mistaken as to what actually qualifies as science. It also shows that Dawkins does not understand Dennett's discussion of actualism in *Darwin's Dangerous Idea,* even though Dawkins extols the work of his fellow bright by calling the book "surpassingly brilliant." How could Darwinism be an "established fact" if actualism demolishes and annihilates it? As we have seen, Dennett tells us that he cannot rule out actualism (although he would like to very much so that he can avoid playing golf). And according to established scientific practice, whatever cannot be ruled out must remain a possibility. We will address Dennett's trenchant analysis in Chapter 8.

Let us now consider Paulos' sexual intercourse calculations (Example 7), a mind-boggling, quantitatively-rich problem that would delight Hugh Hefner. Paulos' views are quite similar to those of Dawkins, and he too confuses knowledge with speculation. Paulos is obviously unaware that his "potential human beings, given all the human ova and sperm there have ever been" is purely hypothetical. Here, "what might have been" takes on the status of a concrete fact. Otherwise, how could he use the term "ipso facto," which means "by the fact itself, by that very fact"? How does a hypothetical speculation become a fact? The answer: it doesn't, unless your reasoning is entirely specious. Paulos can believe whatever he wants, but he is not engaged in science. Considering that Paulos is a renowned mathematician with an impressive pedigree, one might expect more exact and logical thinking. (As I said before, notions of probability and chance are so ingrained into our thinking that most of us are not even aware of their true origin.) Paulos, like Dawkins and others, thinks that those *not* born *could have been* born, hence his use of the word "potential." And those *actually* born *might not have been* born. Talk about the uncertainty! Talk about having your cake and eating it too! Since there are an overwhelmingly large number of "potential" humans that *might have* made it into the world (according to Dawkins and Paulos), this purely speculative notion leads the latter to conclude that those of us who *actually have been* born ("made it to actuality") are "incredibly, improbably fortunate."

We may be fortunate or unfortunate, depending on your viewpoint (consider the innocent person imprisoned on death row or the war-torn refugee in Africa), but our existence has nothing to do with probability or improbability. CFD is omnipresent in Paulos' argument and in all of probability thinking (which has its uses even though it is based on a false assumption). Ian Stewart, writing in *Nature's Numbers*, has noted the tendency of scientists to think about what "could" occur, saying that "there is a philosophical shift from the actual to the potential." I myself find this a rather unfortunate development that is leading us away from reality, not towards it. Alternative scenario thinking, made all the more easy by faster and cheaper computers, isn't leading to an increase in knowledge. Far from it. It is leading us towards an increase in illusion. We can invent and calculate infinite scenarios for both the past and the future, but what did happen and what does happen are about all that matter. Pondering what might happen (and what might have happened) is, of course, endlessly fascinating and keeps armies of people steadily employed, but when the actual is the only possibility, clouds of uncertainty and conjecture evaporate. (I suggest that you read Eliot's *Four Quartets* if you are interested in reflections on the hypothetical and the actual that are far more precise and elegant than those you will find in the confused works of Dawkins or Paulos. Hugh Kenner once said, only half-jokingly, that watching Eliot think would be a fascinating experience in and of itself.)

Paulos speaks of the "fortuitous nature of the world," but this view of reality is entirely subjective and personal (and obviously not open to testing). For Paulos, events are subject to capricious randomness and chance. Presumably he would agree with the standard scientific view that chance made the skies over Paris grey during Henri Becquerel's famous experiments, which led to the "accidental" discovery of radioactivity. Presumably he would agree with the meteorologist's contention that chance events favored the growth of Hurricane Katrina as it intensified over the warm waters of the Gulf of Mexico before slamming into the United States. In Paulos' world, which is in some respects not very different from that of the ancient Greeks, whimsicality and randomness rule. Here, nature is not in order but in a state of disassociation, disintegration, and haphazardness. Anything can happen to anyone at any time.

Paulos maintains that each of us has a "one chance in 68,000 of choking to death; one chance in 75,000 of dying in a bicycle crash; one chance in 20,000 of drowning; and one chance in only 5,300 of dying in a car crash." However, it is false to take observations made over the long run and then spread them over the population at large in order to assign a risk factor to each individual. (We know why Paulos and his colleagues do so, but this doesn't make it any more acceptable.) Only those who will die from a bicycle accident or car crash have a chance of dying, and those who are not going to die have no chance of dying. In the case of a car crash, only one person is at risk; the remaining 5,299 or so people have zero risk. The vast majority of us have nothing to worry about. (No one is going to identify beforehand the person who will actually die, of course.)

Statisticians claim that we live in a world full of hazard, peril, and danger, where practically any of us could be obliterated at any time by some mishap such as a car crash or lightning strike. In a world of "numeracy," chance and probability are said to be lurking everywhere. For example, as a description for his course *"What Are the Chances? Probability Made Clear,"* Michael Starbird tells us that "Life is a matter of probabilities. Every time you choose something to eat, you deal with probable effects on your health. Every time you drive your car, probability gives a small but measurable chance that you will have an accident. Every time you buy a stock, play poker, or make plans based on a weather forecast, you are consigning your fate to probability." Fortunately for us, this brand of so-called "mathematical literacy," which is likely derived from ancient Greek notions of fate and chance, is full of delusion and false perception.

The odds view of reality typified by Dawkins and Paulos is not even wrong, Wolfgang Pauli might venture. As is well known, Pauli reserved this sarcastic viewpoint for scientific theories that were untestable and therefore could not be proven wrong.

It is now time to score the seven scientific selections we have just analyzed. Do they qualify as sound science according to the criteria contained in Sagan's Baloney Detection Kit? Or have our scientists made some major mistakes in logic and reasoning? By the way, sound science does not necessarily mean that a statement is correct, and baloney does not necessarily mean that it is wrong. The purpose of Sagan's test is to determine what qualifies as science, not to determine what is right. I once read that 3,000 astronomers, physicists, and cosmologists were involved in writing a multitude of scientific papers, each of which attempted to explain a highly unexpected astronomical discovery. Even if all the papers qualified as science, there is no way that they could all be right (although they might have all been wrong). I suppose one could argue that each paper could be correct in its own multiverse, but let's not go down that path.

Based on the false logic, sloppy thinking, lack of evidence, and assertions immune to disproof that run throughout our selections to varying degrees, I find myself forced to score the passages as follows:

$$\text{Sound Science} = 0$$
$$\text{Baloney} = 7$$

You can, of course, assign different numbers if you wish. If you are tempted to do so, I suggest you spend some serious time with Sagan's *Demon-Haunted World* and then reconsider.

8

Why Darwinism is Defunct

> My religiosity consists of a humble admiration of the infinitely superior spirit that reveals itself in the little that we can comprehend of the knowable world. That deeply emotional conviction of the presence of a superior reasoning power, which is revealed in the incomprehensible universe, forms my idea of God.
>
> *Albert Einstein, 1927*

I suggested earlier that the question of God is ultimately one of knowledge vs. ignorance, not faith vs. reason. Framing the debate as faith vs. reason implies that the two are mutually exclusive, but it is possible to be reasonable and have faith at the same time. In any case, we can move beyond the framework of faith and reason. We can have knowledge of a designer (God, if you will) through the experience of memory I discussed earlier. We can achieve final answers to the free will problem and know with full certainty that we are here by design. There is no need for unsupported faith. Brights such as Dennett (who announced, "'The time has come for us brights to come out of the closet' and admit publicly that we 'don't believe in ghosts or elves or the Easter Bunny—or God'") demonstrate a surprising, deep, and downright hilarious ignorance.

I have no contribution to make to the Intelligent Design movement here, but it is worth noting that people such as Gould, Dennett, and Dawkins reject God in large part because some living things don't seem to have been designed very well. As Gould maintains in discussing orchids, "If God had designed a beautiful machine to reflect his wisdom and power, surely he would not have used a collection of parts generally fashioned for other purposes. Orchids were not made by an ideal engineer; they are jury-rigged from a limited set of available components.... Odd arrangements and funny solutions are the proof of evolution—paths that a sensible God would never tread but that a natural process, constrained by history, follows perforce."

Gould rejects God because of what he sees as inferior design in nature. Similarly, James Q. Wilson tells us that "if an intelligent designer had created the human eye, He (or She) made some big mistakes. The eye has a blind spot in the middle that reduces the eye's capacity to see." According to Gould and Wilson, if the world were designed by God, things should be more perfect. After all, isn't perfection an attribute we associate with God? What kind of God would be so stupid, so unintelligent?

Whatever the merits of the above arguments, Gould and Wilson judge the existence of God against their own preconceived notions of beauty and ideal design. God doesn't measure up to their standard as a designer, so obviously He doesn't exist. But what if God had intended things to work out this way? What if the odd arrangements and "mistakes" are supposed to be present? What if God wanted to look like a clumsy tinkerer instead of a master craftsman? This is exactly the case, but it is not the heart of my argument, which shares little in common with other anti-

Darwinian views. No, my argument against Darwin and his long list of disciples stems from the certain knowledge that the actual is the only possibility.

Let us turn to Dennett again, an unwitting ally of mine. As he has said, if actualism were true, "Darwinism would be defunct, utterly incapable of explaining any of the (apparent) design in the biosphere. It would be as if you wrote a chess-playing computer program that could play one game by rote (say, Alekhine's moves in the famous Flamberg-Alekhine match in Mannheim in 1914) and, *mirabile dictu*, it regularly won against all competition! This would be a 'pre-established harmony' of miraculous proportions, and would make a mockery of the Darwinian claim to have an explanation of how the 'winning' moves have been found."

Dennett has hit the nail squarely on the head—actualism, which denies CFD, obliterates the idea that evolution could have gone or could go in many different directions. Actualism puts us in the here and now with no ifs, ands, or buts. Actualism demolishes evolutionary biology and vanquishes Dawkins' idea that the thread of your existence is "wincingly tenuous." You can wince no more (if you ever did so in the first place). You also might want to knock on wood to confirm that you are alive at this very moment in time. Actualism makes a virtual mockery of the ubiquitous mantras of adaptation, competitive advantage, and emergence, the main lines of argument taken by those who try to explain why things are as they are. (Please blame Dennett, not me, for bringing the words "defunct" and "mockery" into play, although I must confess to having a great fondness for them. They have an unambiguous finality much like Donald Trump's famous phrase "You're fired" on *The Apprentice*. With "defunct" and "mockery," there is no doubt whatsoever about the infinitely unflattering implications for Darwinian theory. Hey, it's not my fault that nearly all scientists believe in Darwinism, which is neither scientific nor correct.)

Questions such as "what advantage did the survivors possess that the deceased did not?," "why did event A occur and not event B?," and "why is X hardwired into the brain and not Y?," which are found in endless variations in evolutionary thought, are often irrelevant and beside the point (although the explanations they spawn may be quite interesting as well as fodder for never-ending speculation, and, on occasion, extremely nasty intellectual battles. You know the saying—academic feuds are so vicious because the stakes are so low).

For example, Diamond asks why the Europeans conquered the Incas and not the other way around. His reasons such as geographic luck are entirely logical and seem plausible enough, but in the final analysis, luck had nothing to do with it. Similarly, we are asked by others to consider what would have happened if Hitler, or Alexander the Great, or Newton, or some other major historical figure, had not been born. The answers are, of course, by definition entirely theoretical and academic, although they may be quite imaginative. However, I would contend that such theorizing is essentially futile and meaningless. History occurs by design, and whatever happened could not have occurred otherwise. The past could not have been prevented. The reason things are the way they are is that they are *supposed* to be this way. I am not going to go so far as Taleb when he suggests that we "need to downgrade 'soft' areas such as history and social science to a level slightly above aesthetics and entertainment, like butterfly or coin collecting." But it is clear that many of our "explanations" of why things took place the way they did, coming from historians and evolutionary biologists alike, resemble what he calls "narrative dependent studies." Humans seek explanations in order to make sense of the world, but some of our explanations may ultimately prove to be hollow and empty. (See the Notes section for a fuller discussion of Taleb's extraordinarily important work.)

Many from the anti-God contingent wonder how God could allow evil or suffering. They assume that God is benevolent and should not give us a Hitler (or even a destructive tsunami). When they see suffering, they assume that there is no God. As someone who supports Einstein's idea of a God who shows Himself in the harmony of natural law (as opposed to a supernatural God who intervenes in the daily lives of humans), I suggest that things have been designed quite nicely. On a human scale, we feel free and believe that it is in our power to confront evil (e.g., stop a Hitler), but afterwards, we can say that we could not have acted otherwise. I am aware that such a view requires us to fundamentally rethink concepts such as responsibility and morality (as well as guilt, shame, and punishment), but this is a secondary matter. Some of us may want more from a designer, but I think the situation is perfectly fine the way it is.

Inscribed at the entrance to the Oracle of Apollo at Delphi are the famous words "Know Thyself." Let me suggest that following the commandment of knowing oneself leads to the use of memory for liberation, which eventually brings us to God. There is an experience available to all of us that allows us to achieve certainty, peace of mind, and final answers. We can know that we are here by design and that our future is already written. We can know that nature is in a state of law and order. We can know that Einstein was right when he said, "God does not play dice."

Appendix A

Observations on the Direction of Time

Let us briefly address the issue of time and whether or not it has an intrinsic direction. I can by no means give a thorough treatment to the matter here (entire books have been written on the subject), but we can at least put a stake in the ground.

The prevailing scientific view is that the motion of an object such as a billiard ball (or planet) is reversible, whereas the motion of large numbers of billiard balls is irreversible. Time is said to have no preferred direction except in cases involving large numbers not at equilibrium. Eddington puts it this way, "Any change occurring to a body which can be treated as a single unit can be undone. The laws of Nature admit of the undoing as easily as of the doing." Even though he adds, "It may be objected that we have no right to dismiss the starting-off as an inessential part of the problem; it may be as much a part of the coherent scheme of Nature as the laws controlling the subsequent motion...," he maintains, "So long as the earth's motion is treated as an isolated problem, no one would dream of putting into the laws of Nature a clause require that it must go *this* way round and not the opposite" (italics his). Eddington acknowledges that nature may care about the "starting-off" point, but he then claims that there is nothing in nature telling the earth that it must be going in the direction it is actually going.

After a lengthy discussion about the motion of a pendulum and the fact that "the original events" will have shown an actual starting point (from the left or the right), P.C.W. Davies says, "what is important here is not whether the reverse sequence of events *did* happen, but that it *could* happen, being perfectly consistent with the laws of physics and 'everyday' circumstances" (italics his). Notice the use of CFD here; the fact that the reverse sequence is consistent with the laws of physics doesn't lead us to the conclusion that nature has no preference. There is no evidence that a planet or similar object is indifferent as to which direction it is actually going.

In fact, there is a clause in the laws of Nature that is telling the earth to go *this* way (the real way) around the sun. (The real is the only possibility.) We may not be able distinguish a film of the sun's motion running forward (the real world) from one running backward (the unreal world), but this shows our ignorance, not time symmetry on the part of nature. Direction is built into nature; she is not indifferent as to which way something is moving. We may not be able to distinguish a film running forward from one running backward in situations such as a pendulum or billiard ball, but this does not lead to the conclusion that nature is indifferent.

There is actual evidence that nature does have a preferred direction in time and can distinguish between past and future. Greene, Davies, and Martin Gardner are among those who have discussed the CPT (Charge Parity Time) theorem and the fact that the K^o meson decay strongly suggests a preferred orientation in time, but the first two tend to dismiss this as an anomaly having little to do with the real world. My philosophical view is that nature should be consistent throughout (whether at the level of human experience or elementary particles), so if time asymmetry is observed in one place, it should serve as an important clue that this behavior is applicable everywhere. According to Gardner, who discussed CPT symmetry issues raised due to the work of James Cronin and Val Fitch in 1964, "the only way out was to make T asymmetric. Physicists were forced to accept the astonishing fact that on the microlevel, for reasons not yet understood, nature sometimes recognizes a difference between past and future. What Eddington called the 'arrow' of time is actually embedded in certain weak interactions."

More recently, there has been direct evidence for an arrow of time, as CPLEAR (CERN) and KTeV (Fermilab) experiments have shown. Although these experiments are far removed from the world of classical physics and everyday reality, they point the way towards an underlying time asymmetry on the part of nature. One simple and obvious way to achieve overall consistency between the macroscopic and microscopic worlds is by saying that the motion of a billiard ball is irreversible. In other words, nature does care about direction at both the classical and quantum levels, despite our inability in the former case to tell what really took place by looking at the motion of a film in forward and reverse directions. Our lack of knowledge of a ball's starting point is no reason for assuming that it *could have* begun from the right instead of the left.

Appendix B

Einstein, God, and the Number One Musketeer

Einstein's position on God and religion has been studied and debated for decades, with the religious and the atheistic both claiming him as one of their own over the years. During the past summer, I read an ad for Christopher Hitchens' *God is Not Great: How Religion Poisons Everything,* claiming that Hitchens was the Fourth Musketeer. Hitchens had, of course, joined Number One Musketeer Richard Dawkins (*The God Delusion*), Daniel Dennett (*Breaking the Spell: Religion as a Natural Phenomenon*), and Sam Harris (*Letter to a Christian Nation*) as the fourth Big Atheist to have written a bestseller. Victor J. Stenger, who wrote *God: The Failed Hypothesis. How Science Shows that God Does not Exist,* is generally left out of the mix. And for good reason: science does not show that God is absent. (Wasn't it Sagan who was so fond of saying, "Absence of evidence is not evidence of absence"? Just don't hold your breath waiting for ET to phone home.)

This is not a review of the new books on atheism, although I think that many of the arguments presented against God are not particularly original. In addition, the authors tend to unjustifiably lump God and religion together. God can do perfectly well without much of the religion that is out there. What I would like to focus on here is Dawkins' use of Einstein to support the atheistic position. Dawkins explicitly maintains that Einstein was an "atheistic scientist." Dawkins, of course, wants to claim that Einstein sides with him and other proudly atheistic brights. I think a good case can be made that Dawkins is guilty of misappropriating Einstein, much like he accuses the supernaturalists of doing. Referring to the opposition, he says that they are "eager to misunderstand and claim so illustrious a thinker as their own."

It is perhaps worth noting here that I am in full agreement with Dawkins' distinction between Einstein's religion and supernatural religion. Einstein certainly did not believe in a supernatural God, one that intervenes in the daily lives of human beings. Einstein thought that those who held such a belief were rather naive. But this does not mean that he was an atheist, as Dawkins tells us. (Dawkins also says that Einstein showed "pantheist reverence" and contends that "pantheism is sexed-up atheism.")

What did Einstein believe? Here are some of his musings about God and the universe:

> "I am not an atheist. I do not know if I can define myself as a pantheist. The problem involved is too vast for our limited minds."

> "You may call me an agnostic, but I do not share the crusading spirit of the professional atheist whose fervor is mostly due to a painful act of liberation from the fetters of religious indoctrination received in youth."

> "The fanatical atheists are like slaves who are still feeling the weight of their chains which they have thrown off after hard struggle. They are creatures

who—in their grudge against traditional religion as the 'opium of the masses'—cannot hear the music of the spheres."

"Every one who is seriously involved in the pursuit of science becomes convinced that a spirit is manifest in the laws of the Universe—a spirit vastly superior to that of man, and one in the face of which we with our modest powers must feel humble."

"There are people who say there is no God. But what makes me really angry is that they quote me for support of such views."

Reading these passages makes it difficult to view Einstein as an atheist. In the first line of the first selection, Einstein says unambiguously that he is "not an atheist." Although exploring Einstein's concept of God is beyond the scope of this book, it is clear that he accepted the notion of a spirit that is seen in the law and harmony of the universe. He rejected a dice-playing God and notions of inherent randomness, whereas today's scientists tend to accept them unreservedly. It would be difficult to find the concept of "a spirit vastly superior to that of man" among many scientists, judging from the literature. And humility? We've already seen Eddington's unconcealed arrogance regarding the intellectual prowess possessed by scientists ("the attitude is often one of naive surprise that Nature should have hidden her fundamental secret successfully from such powerful intellects as ours"). Spend some time around scientists and look for humility. If your experience is anything like mine, you will have a difficult time.

Let us hear from Walter Isaacson, someone who has probably spent more time researching Einstein than anyone else of late.

In *Einstein: His Life and Universe*, he writes, "Einstein never felt the urge to denigrate those who believe in God; instead, he tended to denigrate atheists.... In fact, Einstein tended to be more critical of the debunkers, who seemed to lack humility or a sense of awe, than of the faithful."

Was Einstein an atheist, as Dawkins claims? You be the judge.

Notes

1. Introduction

p. 1   Albert Einstein, *The Born-Einstein Letters: Correspondence Between Albert Einstein and Max and Hedwig Born from 1916 to 1955 with Commentaries by Max Born*, trans. Irene Born (New York: Walker and Company, 1971), p. 149.

p. 2   Carl Sagan, *The Demon-Haunted World: Science as a Candle in the Dark* (New York: Random House, 1995), p. 28.

p. 5   Nick Herbert, *Quantum Reality: Beyond the New Physics* (Garden City, Anchor Press/Doubleday, 1985), p. xii.

2. Beginning the Journey

p. 12   Werner Heinsenberg, *Physics and Philosophy: The Revolution in Modern Science* (New York: Harper Torchbooks, Harper and Row), see p. 42.

p. 14   Philip J. Davis and Reuben Hersh, *Descartes' Dream: The World According to Mathematics* (San Diego: Harcourt Brace Jovanovich, 1986), p. 18.

p. 15   Ernst Mayr, *What Evolution Is* (New York: Basic Books, 2001), pp. 119–120.

p. 18   Robert H. March, *Physics for Poets* (New York: McGraw Hill Book Company, 1970), see p. 243 for a discussion of Bohr.

3. A Certain World Beyond Science

p. 19 Norman O. Brown, *Love's Body* (New York: Vintage Books, 1966), p. 244.

p. 20 Albert Einstein, *The New Quotable Einstein*, Collected and edited by Alice Calaprice (Princeton: Princeton University Press, 2005), p. 194.

p. 20 Stephen W. Hawking, *A Brief History of Time* (New York: Bantam Books, 1988), p. 175.

p. 21 Michael Shermer, *How We Believe: The Search for God in an Age of Science* (New York: W. H. Freeman, 2000), p. 97.

p. 22 Alfred Jules Ayer, *Language, Truth and Logic*, 2nd Edition (London, 1946; New York: Dover Publications, Inc., 1952), p. 118.

p. 26 Stephen Jay Gould, *The Panda's Thumb: More Reflections in Natural History* (New York: W.W. Norton & Company, 1980), p. 20.

p. 26 Michael Shermer, *Why People Believe Weird Things: Pseudoscience, Superstition, and other Confusions of Our Time* (New York: Owl Books, Henry Holt and Company, 1997), p. xxi.

4. The Solution to the Free Will Problem

p. 27 Walter Isaacson, *Einstein: His Life and Universe* (New York: Simon and Schuster, 2007), p. 392.

p. 29 Daniel Dennett, *Darwin's Dangerous Idea: Evolution and the Meanings of Life* (New York: Simon & Schuster, 1995), see discussion of actualism on pp. 106, 120, 182n, 260.

p. 31 Max Born, *The Born-Einstein Letters,* p. 155.

p. 31   Robert M. Augros and George N. Stanciu, *The New Story of Science: Mind and the Universe* (Chicago: Gateway Editions, 1984), pp. 30–31.

p. 31   Isaacson, *Einstein: His Life and Universe,* see p. 387 and surrounding.

p. 32   Brown, *Love's Body,* pp. 88–89.

p. 33   Dennett, *Darwin's Dangerous Idea,* p. 120.

p. 35   Richard Dawkins, *Unweaving the Rainbow: Science, Delusion, and the Appetite for Wonder* (London: Longmans, 1986), pp. 1–2.

p. 35   Dennett, *Darwin's Dangerous Idea,* p. 260.

p. 38   Richard Feynman, *The Character of Physical Law* (London, 1965; Cambridge, Mass.: MIT Press, 1967), p. 108.

p. 38   Bernard Berofsky, ed. *Free Will and Determinism* (New York: Harper and Row, 1966), pp. 17–19.

p. 39   John Bell, quoted in *The Ghost in the Atom,* edited by P.C.W. Davies and J.R. Brown (Cambridge: Cambridge University Press, 1999), p. 47. (P.C.W. Davies and Paul Davies, author of *The New York Times* op-ed contribution referenced below as well as numerous popular science books, are the same person.) Here is the exchange:

Bell: You know, one of the ways of understanding this business is to say that the world is super-deterministic. That not only is inanimate nature deterministic, but we, the experimenters who imagine we can choose to do one experiment rather than another, are also determined. If so, the difficulty which this experimental result creates disappears.

Editors: *Free will is an illusion – that gets us out of the crisis, does it?*

Bell: That's correct. In the analysis it is assumed that free will is genuine, and as a result of that one finds that the intervention of the experimenter at one point has to have consequences at a remote point, in a way that influences restricted by the finite velocity of light would not permit. If the experimenter is not free to make this intervention, if that is also determined in advance, the difficulty disappears.

p. 39   Feynman, *The Character of Physical Law,* p. 108.

p. 40   Ilya Prigogine and Isabelle Stengers, *Order Out of Chaos: Man's New Dialogue with Nature* (New York: Bantam, 1984), p. xxvii.

p. 42   Einstein, in Calaprice, *The New Quotable Einstein,* pp. 228–229. The two "malicious" quotes are found here.

5. Cracking Nature's Code of Law and Order

p. 45   Einstein, in Calaprice, *The New Quotable Einstein,* p. 196.

p. 45   Steven Weinberg, quoted in *The Atheism Tapes* by Jonathan Miller, BBC, 2004, http//cotimotb.siteburg.com/wiki/index.php?wiki=AtheismTapesTwo.

p. 46   Sir Arthur Eddington, *The Nature of the Physical World* (Cambridge, 1929, rpt. Ann Arbor: University of Michigan Press, 1958), see discussion on p. 74 and following pages.

p. 49   Richard Feynman, Robert B. Leighton, Matthew Sands, *The Feynman Lectures on Physics, Volume 1* (Reading, MA: Addison Wesley Publishing Company, 1963), p. 6-2.

p. 50   Natalie Angier, *The Canon: A Whirligig Tour of the Beautiful Basics of Science* (New York: Houghton Mifflin Company, 2007), p. 48.

p. 51   Richard Tolman, *The Principles of Statistical Mechanics* (Oxford: At the Clarendon Press, 1938), see extensive discussion on pp. 59–66.

p. 53   David Halliday and Robert Resnick, *Physics: Parts I and II, combined 3rd edition* (New York: John Wiley and Sons, 1978), pp. 557–564.

p. 54 G. S. Rushbrooke, *Introduction to Statistical Mechanics*, (Oxford: At the Clarendon Press, 1949), p. 8.

p. 55 The definition of probability as a number from 0 to 1 is from Steps, Probability, http://www.stats.gla.ac.uk/steps/glossary/

p. 55 Herbert, *Quantum Reality*, p. 236.

p. 58 Walter J. Moore, *Physical Chemistry, 4th edition* (Englewood Cliffs, NJ: Prentice-Hall, Inc., 1972), p. 178.

p. 63 Feynman, *The Character of Physical Law*, p. 100.

p. 63 Feynman, *The Character of Physical Law*, p. 122.

6. Quantum Leaps of Faith

p. 65 Albert Einstein, letter to Erwin Schrödinger, 1928, *Letters on Wave Mechanics* (New York: Philosophical Library, 1967), p. 31.

p. 65 Richard Feynman, *The Feynman Lectures on Physics, Volume 1*, p. 6-10.

p. 66 Feynman, *The Character of Physical Law*, p. 146.

p. 66 B. K. Ridley, *Time, Space and Things* (Harmondsworth: Penguin Books, 1976), p. 136.

p. 66 Feynman, *The Character of Physical Law*, p. 145.

p. 68 Einstein, *The Born-Einstein Letters*, p. 91.

p. 68 Tolman, *The Principles of Statistical Mechanics*, see p. 357.

p. 69 Alford Korsybski, see http://alfred-korzybski.mindbit.com/

p. 69 Nathan Spielberg and Bryon D. Anderson, *Seven Ideas That Shook the Universe* (New York: John Wiley & Sons, Inc., 1987), p. 219.

p. 70   Eddington, *The Nature of the Physical World*, p. 179.

p. 70   See John McPhee, *The Control of Nature* (New York: Farrar, Straus and Giroux, 1989).

p. 71   Mayr, *What Evolution Is*, p. 116.

p. 72   Isaacson, *Einstein: His Life and Universe*, p. 332.

p. 72   Stephen Hawking, *Black Holes and Baby Universes* (New York: Bantam Books, 1993), p. 70.

p. 72   Fritz Rorhlich, *"Facing Quantum Mechanical Reality,"* Science 221 (1983): 1251–55.

p. 72   Brian Greene, *"100 Years of Uncertainty,"* *The New York Times,* April 8, 2005.

p. 74   Tolman, *The Principles of Statistical Mechanics,* see discussion on p. 357.

p. 75   See Lee Smolin, *The Trouble With Physics: The Rise of String Theory, the Fall of a Science, and What Comes Next* (Boston: Mariner Books, 2006) and Peter Woit, *Not Even Wrong: The Failure of String Theory and the Continuing Challenge to Unify the Laws of Physics* (London: Jonathan Cape, 2006).

p. 75   Sagan, *The Demon-Haunted World,* see pp. 210–216.

p. 77   Nassim Nicholas Taleb, *The Black Swan: The Impact of the Highly Improbable* (New York: Random House, 2007), p. 48.

p. 80   Paul Davies, *"Taking Science on Faith,"* *The New York Times,* November 24, 2007.

p. 81   Eyvind H. Wichmann, *Quantum Physics, Berkeley Physics Course – Volume 4* (New York: McGraw Hill Book Co., 1967), p. 53.

p. 82   Stephen Hawking, public lecture, *"Does God Play Dice?"* www.hawking.org.uk/lectures/dice.html.

7. Sagan's Baloney Detection Kit Applied

p. 83 Eddington, *The Nature of the Physical World*, p. 179.

p. 84 Sagan, *The Demon-Haunted World*, see pp. 210–216.

p. 84 J. L. Hodges, Jr., David Krech, and Richard S. Crutchfield, *StatLab: An Empirical Introduction to Statistics* (New York, McGraw-Hill Book Co., 1975), p. 11.

p. 85 Martin Goldstein and Inge F. Goldstein, *How We Know: An Exploration of the Scientific Process* (New York: Plenum Press, 1978), p. 323.

p. 85 Gilbert W Castellan, *Physical Chemistry, 2nd edition* (Reading, Mass.: Addison-Wesley Publishing Company, 1971), p. 198.

p. 85 Mayr, *What Evolution Is*, p. 120.

p. 86 Murray Gell-Mann, *Plectics*, essay in *The Third Culture*, by John Brockman (New York: Simon and Schuster, 1995), p. 320.

p. 86 Dawkins, *Unweaving the Rainbow*, p. 73.

p. 86 John Allen Paulos, *Innumeracy: Mathematical Illiteracy and its Consequences* (New York: Hill and Wang, 1988), p. 11.

p. 89 Shermer, *Why People Believe Weird Things*, see pp. xiii–xxvi.

p. 90 Eddington, *The Nature of the Physical World*, p. 72.

p. 91 See Krishna Myneni, "Symmetry Destroyed: The Failure of Parity," http://ccreweb.org/documents/parity/parity.html

p. 91 John Lienhard, "Engines of Our Ingenuity: Chemical Chirality, No. 604," http://www.uh.edu/engines/epi604.htm. Also see Royston M. Roberts, *Serendipity: Accidental Discoveries in Science* (New York: John Wiley & Sons, Inc., 1989), Chapter 12. Roberts discusses thalidomide in a chapter on Pasteur.

p. 93 Herbert, *Quantum Reality*, p. 236.

p. 96   Richard Dawkins, *A Devil's Chaplain: Reflections on Hope, Lies, Science, and Love* (Boston: Houghton Mifflin Company, 2003), pp. 242–248.

p. 98   Richard Dawkins, The World Question Center 2005, www.edge.org.

p. 98   Sagan, *The Demon-Haunted World,* p. 211.

p. 98   Michael Shermer, *Science Friction: Where the Known Meets the Unknown* (New York: Times Books, Henry Holt and Company, 2005), see pp. xii–xviii.

p. 99   Richard Dawkins, in *What We Believe but Cannot Prove: Today's Leading Thinkers on Science in the Age of Certainty*, by John Brockman (New York, Harper Perennial, 2006), p. 9.

p. 101 Ian Stewart, *Nature's Numbers: The Unreal Reality of Mathematics* (New York: Basic Books, 1995), p. 116.

p. 102 Paulos, *Innumeracy,* p. 133.

p. 103 Michael Starbird, description for *"What Are the Chances? Probability Made Clear,"* Course No. 1474, The Great Courses, published by The Teaching Company.

8. Why Darwinism is Defunct

p. 105 Einstein, in Calaprice, The *New Quotable Einstein,* p 195.

p. 105 Daniel Dennett, in *"The Big 'Bright' Brouhaha: An Empirical Study on an Emerging Skeptical Movement,"* by Michael Shermer, *Skeptic,* November 15, 2003.

p. 106 Gould, *The Panda's Thumb,* pp. 20–21.

p. 106 James Q. Wilson, *"Faith in Theory," The Wall Street Journal,* December 24–25, 2005.

p. 107 Dennett, *Darwin's Dangerous Idea,* p. 260.

p. 109 Taleb, *The Black Swan*, p. 171. Earlier, in Chapter Six, The Narrative Fallacy, Taleb writes, "The narrative fallacy addresses our limited ability to look at sequences of facts without weaving an explanation onto them, or, equivalently, forcing a logical link, *an arrow of relationship,* among them. Explanations bind facts together. They make them all the more easily remembered; they help them *make more sense.* Where this propensity can go wrong is when it increases our *impression* of understanding" (italics his).

Taleb's comments bring to mind the famous case of Steven Weinberg and the trimuons. In 1977, Weinberg and colleague Benjamin Lee cancelled a trip to Yosemite so that they could devise a theory that would explain experimental sightings of trimuons that had been rumored. It later turned out that the trimuon effect hadn't been observed at all. One of the major activities of scientists is to devise theories to explain their observations, but in this case, the trimuon sighting was a false positive (using an analogy from medicine). The theory had been created nonetheless. To this day, the trimuons have not been found.

Appendix A
Observations on the Direction of Time

p. 111 Eddington, *The Nature of the Physical World*, p. 65.

p. 112 P.C.W. Davies, *Space and Time in the Modern Universe* (Cambridge: Cambridge University Press, 1977), p. 58.

p. 113 Martin Gardner, *The New Ambidextrous Universe: Symmetry and Asymmetry from Mirror Reflections to Superstrings, Revised Edition* (New York: W. H. Freeman and Company, 1990), p. 244.

Appendix B
Einstein, God, and the Number One Musketeer

p. 116   Richard Dawkins, *The God Delusion* (Boston: Houghton Mifflin Company, 2006), pp. 13, 18.

p. 117   The five passages by Einstein are found in Calaprice, *The New Quotable Einstein*, and Isaacson, *Einstein: His Life and Universe* as follows:

1. Calaprice, p. 196.
2. Isaacson, p. 390.
3. Isaacson, p. 390.
4. Isaacson, p. 388.
5. Isaacson, p. 389.

p. 118   Isaacson, *Einstein: His Life and Universe*, pp. 389–390.

Index

accident
15, 25, 58, 63, 71, 85, 94, 102, 103

actualism
29, 31, 35, 43, 99, 107, 120

anthropic principle
46

Ayer, A. J.
22, 29, 120

Bell, John
38, 39, 121

Berofsky, Bernard
38, 121

Born, Max
1, 31, 67, 119, 120, 123

Brown, Norman O.
10, 19, 32, 120, 121

Castellan, Gilbert
85, 88, 89, 92, 125

certainty
2, 15, 26, 36, 40, 62, 73, 94, 105, 109

CFD (Contra-Factual Definiteness)
4, 55, 59, 60, 72, 78, 79, 80, 93, 94, 95, 99, 101, 107, 112

chance
2, 3, 4, 14, 15, 17, 18, 25, 29, 33, 45, 46, 49, 50, 57, 58, 60, 62, 65, 66, 67, 68, 71, 72, 73, 76, 78, 80, 81, 85, 88, 89, 90, 92, 94, 97, 98, 99, 100, 101, 102, 103

Davies, P.C.W.
39, 79, 112, 121, 124, 127

Dawkins, Richard
15, 34, 35, 86, 94, 95, 96, 97, 98, 99, 100, 101, 103, 106, 107, 115, 116, 118, 121, 125, 126, 128

Dennett, Daniel C.
29, 30, 31, 33, 35, 43, 62, 95, 99, 100, 105, 106, 107, 115, 120, 121, 126

design
3, 15, 25, 26, 42, 62, 63, 64, 71, 97, 98, 105, 106, 107, 108, 110

determinism
29, 37, 39, 45, 65, 67

disorder
14, 46, 47, 48, 49, 53, 54, 58, 59, 60, 61, 62, 90, 94

disorder parameter
54, 55, 59

Eddington, Sir Arthur
  46, 47, 70, 83, 90, 111, 113, 117, 122, 124, 125, 127

Einstein, Albert
  1, 2, 3, 6, 13, 20, 27, 31, 41, 42, 45, 65, 67, 71, 80, 105, 109, 110, 115, 116, 117, 118, 119, 120, 121, 122, 123, 124, 126, 128

Eliot, T. S.
  10, 37, 95, 101

entropy
  46, 47, 48, 53, 54, 55, 56, 60, 61, 90

equally likely
  48, 49, 50, 51, 52, 56, 61, 72

evolution
  3, 15, 29, 42, 63, 98, 106, 107

Feynman, Richard
  38, 39, 49, 52, 63, 65, 66, 90, 121, 122, 123

free will
  3, 27, 29, 30, 31, 38, 39, 42, 62, 63, 105, 121

free will problem
  3, 28, 29, 30, 38, 61, 105

gold mine of consciousness
  i, 21, 22, 31

Gould, Stephen Jay
  26, 35, 106, 120, 126

Greene, Brian
  6, 72, 73, 74, 75, 76, 112, 124

Halliday, Richard
  53, 54, 55, 56, 57, 58, 59, 60, 122

Hawking, Stephen
  6, 20, 72, 81, 82, 120, 124

Herbert, Nick
  5, 55, 77, 93, 119, 123

hero myth
  24

hypothesis of equal *a priori* probabilities, see also equally likely
  4, 49, 56, 88

law of excluded middle
  38

luck
  15, 29, 33, 78, 94, 99, 108

Mayr, Ernst
  15, 71, 85, 92, 119, 124, 125

memory
  31, 32, 36, 37, 45, 58, 68, 105, 109

Mind of God
  i, 20, 40

natural selection
  i, ii, 26, 85, 97, 99

Paulos, John Allen
  86, 100, 101, 102, 103, 125, 126

potential
  17, 20, 35, 59, 61, 68, 79, 86, 94, 95, 96, 99, 100, 101

prediction
 2, 40, 65, 66, 67, 73

probabilities
 4, 49, 56, 58, 66, 68, 69, 74, 76, 88, 102

probability
 2, 4, 5, 14, 48, 49, 50, 52, 55, 57, 59, 60, 61, 62, 65, 66, 67, 69, 71, 72, 76, 77, 85, 87, 88, 89, 97, 100, 101, 102, 103, 123

randomness
 i, ii, iii, 14, 15, 29, 45, 47, 54, 56, 57, 61, 67, 71, 76, 77, 78, 82, 94, 99, 101, 102, 117

Resnick, Robert
 53, 54, 55, 56, 57, 58, 59, 60, 122

Sagan, Carl
 2, 35, 71, 75, 79, 80, 83, 84, 88, 92, 93, 94, 96, 98, 103, 104, 115, 119, 124, 125, 126

scientific method
 2, 8, 13, 17, 20, 22

second law of thermodynamics
 4, 14, 47, 88

Shermer, Michael
 21, 26, 89, 96, 98, 120, 125, 126

Taleb, Nicholas Nassim
 77, 108, 109, 124, 127

Tolman, Richard C.
 51, 52, 68, 74, 88, 122, 123, 124

uncertainty
 11, 12, 31, 32, 33, 36, 42, 54, 65, 67, 68, 70, 80, 81, 100, 101

uncertainty principle
 67, 69

Weinberg, Steven
 20, 45, 122, 127

Acknowledgements

I would like to thank the following people who have been helpful to me at various times over the years: Marshall Cohen, Louis Kampf, Michael Folsom, Huston Smith, Frederick Crews, David Bohm, Karl Popper, Heinz Sternberg, Ginny Sternberg, Charlene Spretnak, Liz Kennedy, Elise Holschuh, Thomas Ray, Jay Abraham, Robert Allen, Ted Nicholas, Jerry Buchanan, Gary Halbert, David Alexander, and Jan Pendzich. I would also like to thank Leonard Klikunas for commenting on earlier drafts of the book and for writing the Foreword. I am also grateful to the Danforth Foundation for its support during my studies at the University of California, Berkeley.